"Professors Louth and Taylor have been at the heart of the UK's independent analysis of defence and security with their work generating a much broader understanding of the complexities of the subject. I commend this book to you."

Lord William Hague, *Chairman of Royal United Services Institute for Defence and Security Studies*

British Defence in the 21st Century

This book analyses UK defence as a complex, interdependent public–private enterprise covering politics, management, society, and technology, as well as the military.

Building upon wide-ranging applied research, with extensive access to ministers, policy makers, senior military commanders, and industrialists, the book characterises British defence as a phenomenon that has endured extensive transformation this century. Looking at the subject afresh as a complex, extended enterprise involving politics, alliances, businesses, skills, economics, military practices, and citizens, the authors profoundly reshape our understanding of 'defence' and how it is to be commissioned and delivered in a world dominated by geopolitical risks and uncertainties. The book makes the case that this new understanding of defence must inevitably lead to new policies and processes to ensure its health and vitality.

This book will be of much interest to students of defence studies, British politics, and military and strategic studies, as well as policy makers and practitioners.

John Louth is Director of Defence, Industries and Society at the Royal United Services Institute, and is also a specialist adviser to the House of Commons Defence Select Committee.

Trevor Taylor is Professorial Research Fellow in Defence Management at the Royal United Services Institute.

Contemporary Security Studies

Series Editors: James Gow and Rachel Kerr
King's College London

This series focuses on new research across the spectrum of international peace and security, in an era where each year throws up multiple examples of conflicts that present new security challenges in the world around them.

Media Strategy and Military Operations in the 21st Century
Mediatizing the Israel Defence Force
Michal Shavit

Ethics, Law and Justifying Targeted Killings
The Obama Administration at War
Jack McDonald

Quasi-State Entities and International Criminal Justice
Legitimising Narratives and Counter-Narratives
Ernst Dijxhoorn

George W. Bush's Foreign Policies
Principles and Pragmatism
Donette Murray, David Brown and Martin A. Smith

Power Relations in the Twenty-First Century
Mapping a Multipolar World?
Edited by Donette Murray and David Brown

Deterring Russia in Europe
Defence Strategies for Neighbouring States
Edited by Nora Vanaga and Toms Rostoks

British Defence in the 21st Century
John Louth and Trevor Taylor

For more information about this series, please visit: www.routledge.com/Contemporary-Security-Studies/book-series/CSS

British Defence in the 21st Century

John Louth and Trevor Taylor

Routledge
Taylor & Francis Group

LONDON AND NEW YORK

First published 2019 by Routledge

2 Park Square, Milton Park, Abingdon, Oxon, OX14 4RN
605 Third Avenue, New York, NY 10017

Routledge is an imprint of the Taylor & Francis Group, an informa business

First issued in paperback 2020

British Library Cataloguing-in-Publication Data
A catalogue record for this book is available from the British Library

Library of Congress Cataloging-in-Publication Data
Names: Louth, John, author. | Taylor, Trevor, 1946– author.
Title: British defence in the 21st century / John Louth and Trevor Taylor.
Other titles: British defence in the twenty-first century.
Description: First edition. | London ; New York, NY : Routledge/Taylor & Francis Group, 2019. | Series: Contemporary security studies | Includes bibliographical references and index.
Identifiers: LCCN 2018024844| ISBN 9781138705029 (hardback) | ISBN 9781315202389 (e-book)
Subjects: LCSH: Great Britain–Military policy. | Industrial policy–Great Britain. | Political planning–Great Britain.
Classification: LCC UA647 .L68 2019 | DDC 355/.033541–dc23
LC record available at https://lccn.loc.gov/2018024844

ISBN: 978-1-138-70502-9 (hbk)
ISBN: 978-0-367-78699-1 (pbk)

Typeset in Times New Roman
by Wearset Ltd, Boldon, Tyne and Wear

This work is dedicated to all those women, men, and their families who work in or support the UK Defence Extended Enterprise. We are grateful for the friendship and wisdom of our colleagues at the Royal United Services Institute (RUSI), especially Dr Lauren Twort, and thank with love our wives for, well, everything.

Contents

Figures

Tables

Boxes

Foreword

The strength of this thoughtful analysis by Professor John Louth and Professor Trevor Taylor, who I've had the pleasure of knowing for a number of years, is that the logic of their thinking on defence in the United Kingdom seems so obvious. As you read their clear-eyed and logical analysis you can be forgiven for forgetting that the approach they adopt is actually fresh and original.

As senior figures at the Royal United Services Institute for Defence and Security Studies, their writing has always been well-informed and has frequently shaped the debate around defence within the UK and beyond. They have been consistent and persuasive advocates of the importance of investment in science and technology to the maintenance of national military advantage. They have also highlighted the importance of skills and competition in the defence industries and articulated the significant part played by the private sector in defence. We may not always have agreed about every detail, but we have at all times shared a conviction about the importance of these issues and a passion to be driven by the evidence. Everyone who cares about the health of the broad defence enterprise should be grateful to them for what they have done, shining a sharp academic yet practical light on important, but sometimes complicated, defence and security ideas and issues.

By considering British defence as an extended enterprise of politics, policy, technologies, industrial capacity, community action, finance, prosperity and, of course, the armed services themselves, they reveal the UK defence system in all of its component – and complex – constituent parts. It is a thought provoking, comprehensive and highly sophisticated approach – and necessary for our understanding if we are to stay safe, at peace, and helping to police the rules-based international order that is central to our prosperity.

This is an important book which I commend to you.

Sir Peter Luff

Minister, Defence Equipment, Support and Technology 2010–2012

1 Introduction

The end of doctrine and loss of military primacy

This book deals with the subject of UK defence in the twenty-first century. For us, the authors, defence is both a national security policy imperative and an interlocking set of strategies, technologies, values, behaviours, skills, relationships, and programmes that constantly evolve and change. It is complicated and involves much more than just the provision, and practices, of men and women in military uniforms – important though this is.

For the casual reader that may seem an odd statement to make. Surely the subject of UK defence is the story of the Royal Navy and Royal Marines, the British Army and the Royal Air Force. What else is there to consider? Well, the answer is, 'a great deal', and this goes to the core argument of this book. Principally, that UK defence has evolved into a complex and multi-faceted public–private, national-international cats-cradle of partnerships and other relationships, offering components that come together to generate defence capabilities, thereby promoting our security in an unpredictable world. This concept of, what we would label, the Defence Extended Enterprise[1] offers citizens, students, commanders, and policy makers alike a substantive opportunity for understanding, influencing, and shaping the critical forces and factors of UK defence. Simply to think of this subject through the lens of the armed forces is to offer an analogue understanding to a post-digital age.

Let us explain a little at the outset. British defence in the twenty-first century has changed profoundly, and not just in terms of the equipment used by the armed forces and the size and shape of the armed services. To illustrate:

- The notable 1998 Strategic Defence Review[2] was led and implemented by a single ministry, whereas in 2015 it was not thought exceptional that the policy review for defence should be nested within a National Security Strategy overseen by a multi-departmental National Security Council.
- In 1996, British defence doctrine was all about operational victory over clearly identifiable enemies. By 2016, British commanders could barely discuss 'victory' as a concept.
- In the First Gulf War in 1991, almost everyone from the West involved in operations in and around Kuwait was in the military. Come the Iraq and Afghanistan conflicts, 2001–2015, at the height of military engagement,

nearly half the people involved in the Western-led coalitions were drawn from the private sector, under contract to the military and different, multiple governments.[3]

Building upon extensive, applied research into the UK defence environment, with extensive access to ministers, policy makers, senior military commanders, and industrialists, this book characterises British defence as a phenomenon that has endured extensive transformation this century. Looking at the subject afresh as this complex Defence Extended Enterprise involving politics, alliances, businesses, skills, economics, military practices, and citizens, our intention is to profoundly reshape the reader's understanding of 'defence' and how it is to be commissioned and delivered in a world dominated by geopolitical risks and uncertainties.

Indeed, when the Conservative–Liberal coalition came to power in 2010, observers of the UK would have been aware of what could be described as a perfect storm of strategic challenges. There was a necessity to address the consuming needs of difficult operations in Afghanistan, resurgent challenges in Iraq and the soon-to-be disruptive explosion of the Libyan collapse and intervention. This was accompanied by a profound decline in government revenues, a rise in borrowing, and a significant gap between Ministry of Defence (MoD) commitments and intentions and the likely income available. Rising powers, both conventional and non-state, were seeking to exploit Western exhaustion with its conflicts of choice and economic shifts were transforming notions of global influence. Throughout this century, the UK has continued to be challenged by profound population movements from the Middle East and Africa to Europe; a resurgent Russia occupying, by force, lands belonging to another state; strategic confusion over what to do about a conflict such as the one in Syria; and reductions in military manpower and recognised gaps in capability. Moreover, as we will discuss, the UK champions the private sector's involvement in traditional defence roles and has outsourced many of its core management capabilities. Anyone returning to working in defence having left the last century would find the present defence ecosystem to be a profoundly different space.

This book describes, explains and critically assesses this transformation, providing an evidence-based analysis of the new public–private and international constructs of defence. It makes the case that this new understanding of defence must inevitably lead to new policies and processes to ensure its health and vitality. Its aim is to become required reading for all students of defence and those already in senior policy-making and managerial roles.

Chapter objectives

By the end of this chapter the reader will understand:

1   How the subject of UK defence has evolved from operations in 1991 to the contingent defence postures of the twenty-first century;

2   Conventional notions of British defence doctrine at the beginning of the epoch and the challenges of twenty-first century military engagements;
3   The idea of UK defence as a complicated extended enterprise involving multiple components;
4   The structure and outline argument of the book as it addresses the themes offered by this introduction.

Chapter structure

We start with the narrative and supporting data relating to British operations in the First Gulf War of 1991. The intention is to provide a baseline of understanding of the practices and effects of a multi-national, alliance-based operation using conventional land, sea, and air forces, as well as elite special forces. This will demonstrate that, within a major international alliance, British policy-making was governmental and the execution of war-fighting and its support, principally, the preserve of the military.

The text moves on to an analysis of British defence doctrine in 1996, based on an MoD publication that states that defence is about military primacy, the maintenance of a strategic and operational aim and, ultimately, victory over a recognised foe. We will test this doctrine against British involvement in the Sierra Leone conflict in 2000 as another example of British policy at the beginning of the century.

The work will then outline the changes in UK defence stances through the Afghanistan, Iraq, and Libyan operations, and the development of a broader public–private partnership for defence – on operations and within the home base. In turn, this will align to the management reforms of the defence procurement organisations and processes from 1998 onwards. The chapter will conclude with a broader explanatory model of UK defence that will form the basis for each subsequent chapter.

First Gulf War: 1991 – military primacy

For most of the post-war period, British security and accompanying defence postures were driven by, and resided within, the country's membership of the North Atlantic Treaty Organisation (NATO), which we come on to discuss in Chapter 2. Yet Britain's major defence operations prior to the 9/11 attack on the United States were the national operation in 1982 to re-take the Falkland Islands, following a surprise invasion by Argentina, and the country's commitment to join the US-led coalition to evict Iraqi forces from Kuwait, which they had occupied in 1990. These pages deal with the latter operation as an example of multi-national military co-operation and accompanying notions of the primacy of the military to defence.

Prelude to war

The Iraqi Army occupied Kuwait in August 1990 in a dispute that had its origins, historically, in Iraqi claims over Kuwait as sovereign territory,[4] and more contemporarily, in Iraq's assertion that Kuwait was regularly exceeding Organisation of Petroleum Exporting Countries (OPEC) quotas for oil production: a manipulation of the OPEC cartel rules that was deeply impacting upon the Iraqi economy.[5] In response to a massive ground invasion by Iraqi forces and the fleeing into exile of the Kuwaiti Emir and government, the United Nations condemned the invasion, passing Resolution 660 demanding an immediate withdrawal of Iraqi troops.[6] Under United States leadership, and subsequent resolutions provided by the UN,[7] Operation Desert Shield (from August 1990 to 17 January 1991) saw the build-up of significant coalition forces in the Gulf region to protect Saudi Arabia and prepare for the liberation of Kuwait. Indeed, it represented the largest build-up of forces for an operation since the end of the Second World War.[8]

Coalition strike

Operation Desert Storm, from 17 January to 28 February 1991 saw the militaries from 34 countries expel by force of arms the Iraqi occupiers from Kuwait.[9] A coalition of just fewer than one million soldiers, sailors, and airmen inflicted approximately 100,000 casualties on the Iraqi forces, destroying almost 4,000 tanks, 110 aircraft and nearly 3,000 artillery pieces. In return, the coalition lost barely 300 personnel, with close to 500 wounded, and with 31 tanks and 75 aircraft destroyed. It was a stunning victory, owing much to superior firepower, a distinct technological advantage, fighting spirit, and the golden nugget of air supremacy.[10]

Importantly, for the British, it was exactly the sort of conflict that their doctrine, training, equipment, and force structures enabled them to contribute towards effectively. The UK committed almost 2,500 tanks, armoured personnel carriers and vehicles to the international force under the British code name Operation Granby. The 1st Armoured Division protected the coalition advance into Iraq and Kuwait, Royal Air Force Tornados participated in 'shock and awe' bombing operations whilst the HMS *Ark Royal* task group deployed to the eastern Mediterranean.[11] In addition to these conventional forces, the British fielded large numbers of Special Forces from the Special Air Service (SAS) and Special Boat Service (SBS).[12] These men raided in small teams acting as forward controllers for air operations, scouting for targets, and mapping enemy deployments. Close to 54,000 British servicemen and women contributed to the operation at the height of the effort to liberate Kuwait, with a similar number held in reserve.

Military primacy

The characteristics of this conflict, on reflection, would contrast profoundly with Britain's future wars of choice in the twenty-first century, in both Iraq (once more) and Afghanistan. Four factors are critical: first, Operation Granby bene-fitted from an unambiguous legal status courtesy of the United Nations reso-lutions to remove Iraqi forces from Kuwaiti soil.[13] Second, the size and breadth of the coalition lent a legitimisation to operations – a unity of world purpose, perhaps – that seems quite a distance from contemporary threats, risks, and poten-tial conflicts. Third, the first Iraq War was a very symmetrical conflict with estab-lished military formations competing with each other using highly conventional tactics and methods – armoured divisions versus armoured divisions; bombing formations verses ground to air ordinance. Lastly, there was a clearly defined aim and identifiable victory point: that being the Iraqi forces significantly degraded and out of Kuwait.[14] Importantly, policy making was governmental, highly ordered, responsive to international alliances, the need for the establishment of a functioning coalition of states, and anchored in the rule of law. Execution, the planning for operations, and the delivery of military effects, was the preserve of the military. As shall be seen, in the years ahead, British commanders and their political masters could only fantasise about such circumstances and certainties.

British defence doctrine and Sierra Leone – military certainty

British defence doctrine – 1996

In the late 1990s, knowledge and understanding of defence doctrine,[15] its application and limits, was thought by the government and military policy makers to provide a common approach to British operations. This provided the basis for a consistent way of thinking about, and planning for, warfare and a method for generating coherent collective action by armed forces that would deliver military success. A country's military doctrine, therefore, captures its core principles in the application of warfare:

> Many countries have adopted Principles of War, focusing on the most important and enduring tenets applicable to the conduct of war derived from experience.... Disregard of them courts failure.[16]

The British principles of warfare, as presented in 1996, are captured in Box 1.1.

At this time, doctrinally, defence was concerned with the role and application of the armed forces and how they contributed to a broader national security stance. As the British described it in 1996:

> Defence is the military contribution to national security and is a major element of a government's wider security policy.... The principal components of defence are the armed forces, an organisation to make decisions and a means of implementing them.[17]

Box 1.1 Principles of warfare – 1996

1   The selection and maintenance of the aim
2   Maintenance of morale
3   Security
4   Surprise
5   Offensive action
6   Concentration of force
7   Economy of effort
8   Flexibility
9   Co-operation
10   Sustainability

Source: British Defence Doctrine, Joint Warfare Publication 0-01
(Ministry of Defence)

As a military imperative, involvement of the private or charitable sectors was not viewed as a function of defence beyond the provision of a national materiel base and infrastructure.

Within these key principles there are one or two important points to make. At the end of the last century, it was deemed critical for every military operation to select and define the aim clearly. This decision, in turn, drives the very manner in which an operation is planned, resourced, and executed.

> The purpose having been defined, it is also fundamentally important that the aim is sustained and that the application of force is matched to the desired end states.[18]

In other words, once politicians and commanders decide on their strategic objectives, the right force numbers with the right equipment levels are committed to that objective and rigour maintained in the delivery of the aim. Critically this is articulated up front and remains unambiguous during the life of the operation. For some commentators, this is the critical failure in British operations since 2001, as we shall come on to discuss.

British doctrine in action – Sierra Leone, 2000

A retired Chief of the Defence Staff, General Sir David Richards and Richard Connaughton have both described the conflict in Sierra Leone that raged in the latter years of the twentieth century as 'pregnant with lessons'.[19] This seems reasonable: whether one is examining the reasons why Sierra Leone descended into the abyss as it did in the 1990s; the role of the Economic Community of West African States (ECOWAS) in nearly bringing order to the country on three occasions; the UN's initial inability to stabilise the country; the role of the British; or the persistent failure of the international community to build on an improving security environment, there is certainly no shortage of relevant topics to study and from which to learn.[20]

The Sierra Leone civil war began in 1991 when a militarised band of men and boys calling itself the Revolutionary United Front (RUF), under the leadership of an ex-army corporal, Foday Sankoh, began to attack villages in eastern Sierra Leone close to the Liberian border.[21] The RUF was astonishingly cruel, with its signature tactic being mass mutilation of the civilian population. An estimated 50,000 people died, and 20,000 more suffered amputations, with machetes primarily used to sever arms, legs, lips and ears.[22] Clearly an act of brutality in and of itself, severing an arm also politically disenfranchised the victim, as Sierra Leone's young democracy required a finger- or thumb-press on the ballot sheet for an individual to exercise his or her right to vote.[23]

Throughout 1991 and 1992 the RUF conquered much of Sierra Leone, securing control of the country's one key economic asset: its diamond mines in the eastern Kono District.[24] Its operational concept was to clear the land of any potential opposition by destroying villages and towns, killing the residents or undertaking group amputations as a warning to other members of the populace. Young boys were routinely captured and viciously trained as child soldiers, controlled through a cocktail of drugs, industrial-strength alcohol, and fear.[25] These victims became brutalised killers themselves and an aggressive and unpredictable opponent to British forces when the latter deployed as peace-makers and peacekeepers in 2000.

Charles Taylor, the president of Liberia from 1997 to 2003, was convicted for crimes against humanity, in April 2012, at the special court in The Hague, for his role in directing and supplying the soldiers of the RUF.[26] In his summing-up, the presiding judge, Richard Lussick, described how more than 1,000 children had the letters 'RUF' carved into their backs as signs of ownership and fealty.[27]

Taylor stepped down from the Liberian presidency in 2003 to avoid prosecution and was offered a home in Nigeria, having been indicted by the special court established to address crimes committed in Sierra Leone. Under pressure from the international community in 2006, the Nigerians allowed Taylor to be arrested as he tried to leave the country, from where he was delivered to The Hague for trial. At the height of his crimes, it is estimated that Taylor had personal assets of at least $400 million, although by 2012 only about $7 million had been recovered.

The next nine years from 1991 saw a fluid, ruinous, and cruel civil war leading to the Lomé agreement of July 1999, which established a form of power-sharing between the government of Sierra Leone, led by the elected Ahmad Tejan Kabbah, and the RUF led by Sankoh.[28] However, in truth, there was little of Sierra Leone left to govern.[29] To the surprise of very few people in Sierra Leone, in 2000 power-sharing collapsed. Rather, the institutions of the international community that had been sent to assist redevelopment – especially the United Nations Mission in Sierra Leone (UNAMSIL) – found themselves at the mercy of the RUF, which was marching murderously towards the capital, Freetown.[30] The UN, powerless to stop the RUF, started to evacuate their civilian staff from Sierra Leone as both the mission and the government appeared on the verge of collapse.[31]

At the request of the UN and the Sierra Leone government, the British then stepped into the breach with, initially, a non-combat evacuation mission. This comprised a fighting regiment, initially 1 Parachute Regiment, replaced by 42 Commando, Royal Marines, a maritime task force with air power provided by Sea Harrier aircraft, and RAF Chinook and RN Sea King support helicopters.

The initial mission was to evacuate British and friendly civilians in the face of the RUF onslaught towards Freetown. In a change of strategic aim, that quickly turned into combat and mentoring operations to secure the Freetown peninsular, defeat the RUF, and generate new resolve within the UN mission.[32] When the British actually engaged the RUF in battle, the British Army's overwhelming tactics, training, and firepower left the insurgents in little doubt that a new military power now dominated the ground in Sierra Leone. British doctrine of the maintenance of the aim – victory over the RUF – and superior, sustained firepower – the concentration of force – ensured British success.

In fact, being the de facto power in the land allowed the British Army, and the British commander, Brigadier David Richards, to stretch its mandate to the limit.[33] By securing the ground around Freetown, Richards allowed the UN to regroup and reorganise, thereby buying the Sierra Leone government precious time. The RUF splintered and fragmented in the face of British military resolve, with many RUF leaders detained, including Sankoh himself. This was the beginning of the process that led to a ceasefire in November 2000, which finally signalled an end to the conflict and, more importantly perhaps, an opportunity to rebuild a shattered country. The application of British defence principles and capabilities proved the winning formula.

Twenty-first-century conflicts of choice – military questioning

9/11

In the twenty-first century, one of the seminal acts to shape future defence and security postures and behaviours was the attack on the United States by the Al-Qaida terrorist network on 11 September 2001. In brutal suicide missions, civilian passenger aircraft were smashed into the twin towers of the World Trade Center in New York and into the Pentagon. The UK political response was immediate and unambiguous: to pledge fealty to the United States and to join them in the intention to prevent the use of the ungoverned spaces of Afghanistan as a safe haven and base for the planning and mounting of any future terrorist operations. In 2002, a new chapter was published as an addendum to the 1998 Strategic Defence Review, which made clear that British defence policy was to aggressively interdict terrorism at home by so-called 'forward defence' in Afghanistan or anywhere else that an identified or supposed threat was to emanate from.[34] In short, British policy at the start of this century was conformity, supine or otherwise, to US policy and represented:

> An approach informed by the vacuous idea of a global war on terrorism: a war against a condition and therefore, by definition, unwinnable and with few pointers to the appropriate application of the instruments of national power.[35]

Britain's smart, rational, pragmatic defence doctrine, as evidenced, for example, by the First Gulf War and operations in Sierra Leone, had given way to knee-jerk responses and glib phraseology.

Moreover, NATO's invocation of Article V of its founding treaty, whereby all member nations respond to an attack on one,[36] gave credence to an inalienable responsibility to respond through military force. That response was an invasion to remove the Taliban government and the elimination of its Al-Qa'ida partner. In 2002, the War on Terror would shift its attention to Iraq, with British policy developed to join a subsequent US-led invasion of that country in 2003, citing as motivation that Iraq harboured weapons of mass destruction – an intelligence-led accusation that would prove a falsehood.[37]

Consequently, within a few short years British defence doctrine and practice morphed from the certainties associated with conventional state-on-state conflict associated with the First Gulf War and the policy stance of the late 1990s to contingent, highly controversial operations in Iraq and Afghanistan. The wars, and continuing paths of post-conflict recovery and bitter insurgencies, in these countries are treated elsewhere and will not detain us here, other than to note that the loss of certainty within defence policy making afforded by these operations is hugely significant. Just how much so will be discussed in the following chapters.

Defence acquisition – the championing of change and management

In addition to the operational aspects of defence, the pursuit of effective equipment for the use of the British military is as old as warfare. Towards the end of the twentieth century, acquisition programmes for defence capabilities were often reported as being unfit for purpose, inappropriately conceptualised and designed, overtaken by the speed of technological development, late, and budget-busting.[38] The MoD and single services were perceived to lack expertise in capability management, programme and project management, financial management, and risk management. Across these topics the common denominator was the focus, of course, on the notion of 'management'.[39] There seemed to be no clear sense of who the customer actually was with the MoD, and poor delegation of lines of responsibility and financial control. Money appeared to be spent by a single department, the Ministry of Defence Procurement Executive (MoDPE), in a manner that was perceived as inefficient, inconsistent, and possibly undemocratic.[40]

In response to these perceived failings, the MoD was charged by the government of John Major, in the 1990s, with considering significant changes to what was sensed to be a failing defence procurement process and ineffective MoD organisational structure. In parallel, the Labour Party in opposition was developing its own defence review, which focused on a Smart Procurement Initiative as an element of an all-embracing Strategic Defence Review. Thus, following a Labour Party election victory in 1997, the reforms associated with Smart Procurement, rebranded as Smart Acquisition, were rolled-out across the defence environment in the UK from 1998 onwards. As we come on to discuss in the following chapters, these reforms were essentially *structural* and *procedural* in nature.

Smart Acquisition structures

The MoDPE was scrapped and the Smart Acquisition reforms introduced an organisational customer within the MoD itself. The Defence Equipment Customer (DEC) was created with terms of reference to identify the equipment needs for future project teams to successfully procure against. The research, development, and purchase funds would flow from the DEC to the project team and, thereafter, through commercial contracts to industry. Also, a formal equipment supplier, integrating all appropriate project processes and competencies, was formed through a new organisation called the Defence Procurement Agency (DPA). Throughout the DPA, integrated project teams (IPTs) from the military, civil service, and industry were tasked with delivering for the DEC a unique set of co-ordinated activities, equipment, and support contracts to generate an identified military capability within defined and scheduled start and finishing points. These were to be undertaken – usually by industry – within agreed time, cost, performance, and integration parameters, paid for by the DEC.

A large logistics support community was created to manage equipment after its purchase and delivery date. This community was called the Defence Logistics Organisation (DLO) and was charged with developing a unique relationship with the armed forces to secure and enable the effective ongoing maintenance of equipment and future upgrades, largely through support contracts with industry.

A high-level process for all defence procurement activities, known as the CADMID Cycle (comprised of the first letters of Concept, Assessment, Demonstration, Manufacture, In-service, Disposal), was mandated, to which all equipment projects had to conform.[41] Within this process, key milestones for financial approval were set, with this approval coming from the MoD's most senior staff sitting as an Investment Approvals Board.

In addition, Smart Acquisition sought a re-defined, energised, and engaged UK defence and security industrial base, fully integrated within this change programme. A Defence Industries Council was formed under the Blair government, after 1997, comprising senior figures from the MoD and the UK's leading defence and security industry actors. A Ministerial Steering Group, chaired by a defence minister, was established to oversee these arrangements and the relationships, it was hoped, that would emerge.

Last, the MoD sought through the Smart Acquisition change initiative to insert perceived modern skills and competencies within military, civil service, and industrial staff. This intent was pursued through the development of the Acquisition Leadership Development Scheme and the Acquisition Stream,[42] both of which were new and centrally funded national development and learning initiatives.

Within this model, Smart Acquisition possessed a number of key, purposefully designed features. First, it sought to generate what practitioners described as 'a whole-life approach'[43] to defence procurement, embodied within a single IPT, accountable to both the Defence Procurement Agency and the Defence Logistics Organisation. In 2007, the functional divide between the two organisations was itself perceived as too cumbersome and inflexible, and both organisations were combined into the Defence Equipment and Support (DE&S) organisation. The project teams, however, retained their focus as the guardians of the newly procured equipment for the life of the capability, physically transferring from procurement to support at a pre-considered level of programme maturity. Critically, from inception, industry was perceived as an intrinsic part of these project teams.

The DEC provided a clearly identified customer for individual project outputs. As an organisation it was designed to take responsibility for identifying the capability required to meet the UK's military and defence objectives, for translating those requirements into an approved equipment programme and for acting as the organisational customer for the equipment until it entered service. Thereafter, a specific branch of the armed forces was to take responsibility for converting the procured equipment into a usable military capability, managing the equipment in-service, and providing relevant and timely expertise to support the DEC's search for future defence capability.[44]

The Smart Acquisition model assumed a willingness to identify, evaluate, and implement effective trade-offs between equipment performance, their whole-life costs, an annualised cost of ownership to the MoD, and, of course, delivery times and project delays. Moreover, Smart Acquisition was said to sit upon an open and constructive relationship with industry, based on partnering principles and the identification of common goals underpinned by competitive contractor selection whenever this provided the best value for money.[45]

It seems striking that this change programme was framed and articulated in the language of business and management rather than notions of public service, be they military or civil, or extant defence doctrine. This is of profound significance, as we discuss in subsequent chapters. For the moment, however, these features remained wedded to self-described core values and beliefs that, under Smart Acquisition, defence procurement practitioners within the military, civil service, and industry were said to embrace. These core values were:

- An empathy with the customer … a commitment to providing equipment, which meets the user's needs, on time and budget.
- The drive to deliver a high level of performance.
- A desire to work co-operatively with fellow team members and others, valuing the diversity of the team and understanding the different role of colleagues.

- A predisposition to share ideas and information, and the resolve to overcome problems.
- A wish to challenge convention and improve processes rather than hide behind 'the rules' and be satisfied with current performance norms.[46]

During the early years of the twenty-first century, the MoD's developing, ongoing guidance on Smart Acquisition asserted that the change programme had evolved and matured significantly beyond its early promise.[47] With the high level model of Smart Acquisition in place and with a legacy of self-assessed success, the change initiative would direct energies towards an enduring theme of a through-life equipment capability to reduce whole life costs and project time-scales at lower levels of risk.[48] In addition, for the first time, Smart Acquisition sought to cover the provision of all defence requirements, not just those that were equipment-based, but also support services such as guarding, real estate, and business information systems. Its champions were confident that it could cover both conventional activities and contracts and those outsourced through public–private partnerships.

This is a remarkable managerialist vision for Smart Acquisition and broader defence reform, breath-taking in its range and scope of services: there is no offensive or defensive activity within UK military operations that has not been enfolded into the Smart Acquisition process, stated behaviours, desired competencies, or managerial language. Indeed, the 2004 version of the Smart Acquisition guidance overtly sets as its goal the application of Smart Acquisition principles throughout the whole of the MoD and UK defence.[49]

Defence as multiple components

This short introductory chapter displays a very clear historical line in relation to the development of understandings of UK defence in the twenty-first century. In 1991, with the First Gulf War, the British defence effort was, essentially, the projection of kinetic force of arms through the capabilities and bravery of military men and women. Utilising thought-through and considered defence doctrine and postures gleaned from the Cold War and operational experiences in Northern Ireland, the British were able to maintain focus around victory in places such as Sierra Leone in 2000.

This confidence and doctrinal certainty seemed to desert the British in wars of choice in Iraq, Afghanistan, and Libya during the early years of this century. Coupled with continuing managerial reform and privatisation, initially focused on the complexities of acquisition and support, the subject of defence has morphed into a complicated set of factors, forms, features, and processes that we come to discuss in the chapters ahead, through the multiple lenses of the model presented in Figure 1.1.

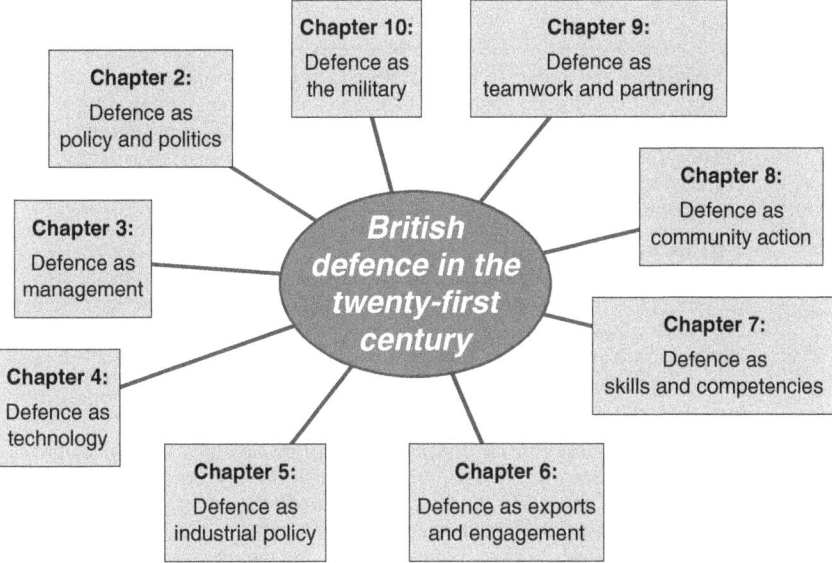

Figure 1.1 UK defence analysis in the twenty-first century.
Source: authors.

As a consequence, the following chapters are structured as described.

Synopsis and chapter outline

Chapter	Title
2	**Defence as policy and politics** This chapter will trace the evolution of UK policy, demonstrating the emergence of defence as a sub-set of security policy, but also highlighting the continuing ambition of successive administrations that UK defence capabilities should be a key element in the country possessing 'global reach' and being 'strong, influential and global'.[50] A capacity for a national action and stance is seen as particularly important in the nuclear deterrent domain. Yet government policy also seems to imply that the UK will never again undertake an operation without state partners. What are the key allies and alliances needed for this policy stance to be relevant? Within the context of defence as politics, the chapter considers the amount of our national treasure committed to defence and unpacks its allocation across the public and private sectors. The work will also consider political decision-making in the context of operations and discuss the lessons learnt from such practices. This analysis will be supported by discussions with serving and retired military leaders and other ranks.

continued

Synopsis and chapter outline *continued*

Chapter	Title

We also track the impact of political decision-making in the twenty-first century on the ability to generate sustainable military capabilities. This will include an analysis of the drive towards privatisations of key assets and the opportunities, risks and uncertainties this contains. The analysis will be enriched by comments drawn from interviews with past defence secretaries, procurement ministers, and opposition spokesmen and women.

3   **Defence as management**
Building upon understandings of complexity and enterprise awareness, this chapter posits that defence is just another management challenge, residing in a policy narrative that champions the private over the public. The chapter explores the government's change management agenda for defence and unpacks traditional notions of certain defence practices as 'inherently governmental'.

The chapter explores the key managerial decisions around procurement and future platforms, revealing a confused picture of the UK's order of battle and its inventory of assets and consumables. It introduces the idea that there is a disconnect between the management of the procurement of capital assets, such as the Queen Elizabeth class aircraft carriers, and the recruitment and retention of skilled people to serve on them.

Within this analysis, the key themes of defence efficiency, effectiveness, and economy will be explored, along with the notion of value for money in defence. Practices prevalent across defence, such as risk and opportunity management and earned value management, will be defined and reviewed to assess their utility. Moreover, the organisational reforms within the MoD, the armed forces and the procurement organisation themselves will be explored as exemplars of this new public management.

4   **Defence as technology**
For people in the West, defence has traditionally been about high-tech capabilities of varying quality and maturity. The US and British stance, traditionally, has been to conduct operations through technological superiority rather than numerical advantage. However, near-peer rivals such as China and Russia are fast catching-up with US and, to a lesser extent, UK technological capabilities. This chapter explores what this means for the level of investment needed for research and development and how this is currently being manifested.

The chapter will explore key emerging technologies, US strategies to harness them and UK efforts to stay relevant within an increasingly competitive sector. It will reflect the increasing trend for husbandry of technologies within commerce rather than government, and ask what this means for the development of enduring and battle-winning capabilities.

5   **Defence as industrial policy**
The text will explore the tensions between stated acquisition policies and de facto policies and practices that have led to a 'pick-and-mix' defence economic landscape. Within this analysis we consider the UK's ambitions for technological advantage on the battlefield, applied research and the broader management of science and technology within a notional UK defence industrial policy.

Issues of intellectual property, transfer policies, ownership, and national independence will be explored within this chapter. Notions of global interdependencies will be offered to challenge conventional explanations of national sovereignty and singular national interest. Corporate strategies and policies will be reviewed to see whether they align with emerging national strategies.

6   **Defence as exports and engagement**
As well as keeping the British people safe, UK defence is now charged by government with contributing to British prosperity through defence exports and sales. This chapter explores what this means for the development of national capabilities and the policies and practices necessary to generate successful export campaigns. It asks whether the twin imperatives of national capability generation and international sales are coherent and realisable.

The text explores the profound organisational effort and initiative from across government and industry in the UK to create the conditions for defence innovation and exports. It questions why innovation is seen as significant for exports rather than home consumption, given that the UK relies upon the maintenance of technological advantage over potential foes. Should new and innovative capabilities find a home with the British forces before they are offered elsewhere? What does this mean for corporate investment profiles?

7   **Defence as skills and competencies**
Much work has been undertaken in the Ministry of Defence since 2010 to define and develop the appropriate skills and competencies across the broader defence enterprise. As discussed, defence policy in the UK is concerned with the generation of force to defeat foes through the application of technological advantage. This requires a highly skilled and competent workforce.

The chapter considers where these supposed skills now reside, government or industry, and what happens when skills and jobs leak from the defence sector. This builds on empirical data gathered to explore the withering of the defence skills base. The work explores the reviews undertaken in MoD and industry on the maintenance and maturation of the defence workforce.

8   **Defence as community action**
A profound, observable, change in the UK, in tandem with the Iraq and Afghanistan conflicts, has been the growth of the defence charitable sector. Based on engagement with policy makers and practitioners, this chapter explores the expansion of this area, its implicit criticism of the Ministry of Defence and wider government, and the consequences on post-conflict provision for the injured and bereaved.

Within this chapter we also explore the role and purpose behind the Armed Forces Covenant[51] and its contribution to the UK defence sector following legislation by the coalition government at the beginning of the decade. We consider the views of the many industrialists and community leaders who have committed their organisations to the Armed Forces Covenant and question whether this community of the willing is being effectively utilised by the MoD. The chapter explores the impact of so many corporate and community pledges under the auspices of the Covenant and what this phenomenon represents for the notion of military exceptionalism.

continued

Synopsis and chapter outline *continued*

Chapter	Title

9   **Defence as teamwork and partnering**
The relationships that are supposedly needed, including so-called 'partnerships,' extend beyond borders and sectors deep into international corporations. This chapter explores the interdependencies of these new and emerging partnerships and seeks to understand the nature of a defence practice now delivered through myriad organisations with multiple societal and commercial purposes.

This is especially significant as much of the changes to defence in the UK this century have been couched in a managerial taxonomy and discourse of partnership and partnering, invariably with the private sector seen as the more energetic and capable bed-fellow. This chapter frames the narrative to allow the question of where the military now feature – which is explored in the following chapter.

10   **Defence as the military**
When defence is conceptualised as a range of competing perspectives, such as those above, the reader will be asked to reflect upon where this leaves the armed forces. Having introduced organisational reform of the military, this chapter will explore the revised purpose of the British armed forces, working in strategic and contingent partnerships with the private sector, other nations and ad hoc companions possessing short-term shared interests. The chapter will explore how this broader enterprise approach to defence retains the military at its centre and what this means for doctrine, training and development.

At the heart of this will be a critical analysis of the use of reserves and sponsored reserves and what this represents in practice for the broader economy and society in the UK. The chapter will reflect upon traditional notions of military service and consider whether these are relevant and appropriate today.

Lastly, the chapter will attempt to align British forces and capabilities to current notions of threats and conflict. Given the profound changes – deliberate and accidental – evidenced by the previous chapters, it would be a shock for Britain not to possess the most appropriate capabilities available for the modern epoch. Do the armed forces and their newly strategic partners have the right numbers, equipment, support, techniques, training, and experience for success? Who makes this judgement and how can the public have confidence that it is right?

11   **Conclusion: defence practice – from analogue to digital? The Defence Extended Enterprise**
The authors will re-frame defence from its conventional geopolitical and military explanations into a complex portfolio of actors and influences that are described as the Defence Extended Enterprise. This concept is explored through the theoretical literature on enterprise systems supplemented by interviews with senior politicians and commanders charged with preparing the military in 2016.

The chapter brings together the multiple roles and practices of the defence industrialist, technologist, and myriad supply chains. It positions defence management as the manipulation of an ever more complicated national and international knowledge economy.

> We are told that today's forces are more capable, possess greater technologies within their platforms and weapons, are better trained, and have greater access to expertise than any previous generation. It is almost as if our defence forces have had their 'analogue to digital' moment. Indeed, the capabilities of British systems in this decade are greater than those available to those involved in, say, the Korean War. But does that make us safer and do we have the right forces for today's threat profile and policy stance?
>
> The earlier chapters suggest that the Defence Enterprise has to be explained through multiple building blocks of capability more complex than at any previous time. Very few people – politicians, military leaders, chief executives – are prepared to see defence in this way and prefer the certainties of yesterday's military-centric defence explanation. Only when we understand the system that we have created in the UK can we take a judgement on how effective, efficient, and fit for purpose it is.

Conclusion

This chapter has explored the manner in which the nature of defence in the UK has evolved from the doctrinal certainties of the Cold War, and its immediate aftermath, to the complexities of contingent operations today. The clarity of defence doctrine in the 1990s has given way to responses to profound shocks such as the 9/11 terrorist attacks and the challenges – not to say, failures – of our engagements in Afghanistan and Iraq. We have introduced the idea, expanded upon in the following chapters, that UK defence must be characterised as a complicated extended enterprise involving multiple components and building blocks, some in government and others in the private sector; some on shore, others drawn from international partners. We now turn to defence as policy.

Notes

1 The Defence Extended Enterprise includes organisations, people and competencies that allow for the generation and use of national defence capabilities. This concept goes beyond the MoD to include the armed forces, broader governmental bodies, defence manufacturers, and suppliers. See, for instance, John Louth and Trevor Taylor, 'Beyond the Whole Force: The Concept of the Defence Extended Enterprise and its Implications for the Ministry of Defence', *RUSI Occasional Paper*, 2015.

2 Ministry of Defence, *Strategic Defence Review: Modern Forces for the Modern World* (London: The Stationary Office, 1998).

3 John Louth and Peter Quentin, 'Making the Whole Force Concept a Reality', *RUSI Briefing Paper*, 2014, p. 4.

4 See, for instance, Richard Alan Schwartz, *Encyclopedia of the Persian Gulf War* (North Carolina: MacFarland, 2008), p. 45.

5 See, for instance, Ewan W Anderson and Liam D Anderson, *An Atlas of Middle Eastern Affairs* (London: Routledge, 2014), p. 221.

6 United Nations Security Council, Resolution (UNSCR) 660, August 1990, https://documents-dds-ny.un.org/doc/RESOLUTION/GEN/NR0/575/10/IMG/NR057510.pdf?OpenElement, accessed 8 November 2017.

7 A number of other Security Council Resolutions and Arab League resolutions were used to encourage Iraqi troops to withdraw from Kuwait including economic sanctions and reaffirming the independence and territorial integrity of Kuwait. Resolution 678 provided Iraq with a withdrawal deadline, failure to comply allowed for the legal use of force.

8 Yigal Sheffy, 'The Military Dimension of the Gulf War', in Ami Ayalon (ed.), *Middle East Contemporary Survey,* Volume XV: 1991 (Oxford: The Moshe Dayan Center, 1993).

9 Ibid., p. 6.

10 See, for instance, Anthony Tucker-Jones, *The Gulf War: Operation Desert Storm 1990–1991* (Barnsley: Pen and Sword, 2014).

11 Theo Farrell, Sten Rynning and Terry Terriff, *Transforming Military Power since the Cold War: Britain, France, and the United States, 1991–2012* (Cambridge: Cambridge University Press, 2013), p. 134; Anthony H Cordesman, *The Iraq War: Strategy, Tactics, and Military Lessons* (Washington DC: CSIS, 2003).

12 Tucker-Jones (2014), op. cit., p. 89.

13 This includes UNSCR 660, 661, 662, 664, 665, 666, 667, 669, 670, 674, 677, 678.

14 The objectives of Operation Granby were described by the MoD in a supplement to the London Gazette 29 June 1991, 'The initial objective of Operation GRANBY was to help deter any further aggression by Iraq in the Gulf and particularly against Saudi Arabia. The objectives were later expanded to: secure, together with our Coalition allies, a complete and unconditional Iraqi withdrawal from Kuwait; restore the legitimate government of that country; re-establish peace and security in the area; uphold the authority of the United Nations', United Nations, Supplement to the London Gazette, 29 June 1991 www.thegazette.co.uk/London/issue/52589/supplement/38/data.pdf, accessed 8 November 2017.

15 Doctrine can be characterised as 'that which is taught and valued'. A better definition for military doctrine is 'the fundamental principles by which military forces guide their actions in support of objectives. It is authoritative, but requires judgement in application.' This definition could be found in NATOS's Glossary of Terms and Definitions, 1996.

16 Ministry of Defence, Joint Warfare Publication (JWP) 0–01, *British Defence Doctrine*, (London: MoD, 2006) p. 1.4.

17 *Ibid.*, pp. 3.4–5.

18 *Ibid.*, p. Annex A.2.

19 See, for instance, Richard Connaughton, 'Military Intervention and Peace-keeping: The Reality', in *Joint Forces Quarterly Review*, 26 July 2001. See also Gwyn Prins, *The Heart of War* (London: Routledge, 2002).

20 David Richards, 'Sierra Leone: Pregnant with Lessons', in David Richards and Greg Mills (eds), *Victory Among People: Lessons from Countering Insurgency and Stabilising Fragile States* (London: RUSI, 2011).

21 David Keen, *Conflict & Collusion in Sierra Leone* (Oxford: James Currey Publishers, 2005), p. 36.

22 Udy Bell, *Sierra Leone: Building on a Hard-Won Peace* (UN Chronicle No. 4, 2000).

23 John-Peter Pham, *Child Soldiers, Adult Interests: The Global Dimensions of the Sierra Leonean Tragedy* (New York: Nova Publishers, 2005), p. 115.

24 See, for instance, Lansana Gberie, *A Dirty War in West Africa: The RUF and the Destruction of Sierra Leone* (Indiana: Indiana University Press, 2005).

25 See, for instance, Myriam Denov, (2010) *Child Soldiers: Sierra Leone's Revolutionary United Front* (Cambridge: Cambridge University Press, 2010).

26 For more information on the trial of Charles Taylor and other individuals from the Sierra Leonean civil war who have been tried visit the International Justice Monitor website, a project of the Open Society Justice Initiative, www.ijmonitor.org/charles-taylor-background/#fourteen, accessed 2 October 2017.

27 Owen Bowcott, 'Charles Taylor aided and abetted Sierra Leone war crimes, Hague court finds', *Guardian*, 26 April 2012.

28 For more information on the content of the Lomé agreement see, Government of Sierra Leone (1999) Lomé Peace Agreement. Freetown, Government of the Republic of Sierra Leone, www.sierra-leone.org/lomeaccord.html, accessed 10 August 2017.

29 See, *The Economist*, 13 May 2000.

30 Gberie (2005), op. cit.

31 'Funmi Olonisakin, 'United Nations Mission in Sierra Leone (UNAMSIL)', in Joachim Koops, Joachim Alexander Koops, Norrie MacQueen, Thierry Tardy, Paul D Williams (eds), *The Oxford Handbook of United Nations Peacekeeping Operations* (Cambridge: Cambridge University Press, 2015).

32 One of the authors, John Louth, was a serving British officer at the time and served in Sierra Leone during the combat phase of the operation.

33 Larry J Woods, *Military Interventions in Sierra Leone: Lessons from a Failed State*. (Kansas: DIANE, 2010).

34 'The Strategic Defence Review: A New Chapter' can be accessed online at: http://archives.livreblancdefenseetsecurite.gouv.fr/2008/IMG/pdf/sdr_a_new_chapter_cm5566_vol.1.pdf, accessed 9 August 2017.

35 See Robert Fry and Desmond Bowen, 'UK national Strategy and Helmand', in Michael Clarke (ed.), *The Afghan Papers: Committing Britain to War in Helmand, 2005–6* (London: RUSI, 2011).

36 Commonly referred to as collective defence.

37 See, for instance, Sir J Chilcott, 'The Report of the Iraq Inquiry', House of Commons Report HC264, 2016.

38 See, for instance, Bill Kincaid, *Dancing with the Dinosaur: How to do Business with the MoD in the Smart Procurement World* (Newcastle-upon-Tyne: UK Defence Forum, 1999).

39 *Ibid.*

40 John Louth, *A Low Dishonest Decade: Smart Acquisition and Defence Procurement into the New Millennium* (Cardiff: UWIC, 2010).

41 Ministry of Defence, *The Smart Acquisition Handbook* (Edition 4), (London: MoD, 2002), p. 4.

42 See, for instance, John Louth, 'Defence Acquisition Reform and the British Condition Promises, Betrayal and Resignation', in Kevin Burgess and Peter Antill (eds), *Emerging Strategies in Defense Acquisitions and Military Procurement* (Hershey, US: IGI Global, 2016).

43 Includes the whole life cost and cost of ownership which spans the CADMID cycle.

44 MoD (2002), op. cit., p. 9.

45 *Ibid.*, p. 25.

46 *Ibid.*, p. 3.

47 Ministry of Defence, *The Smart Acquisition Handbook* (Edition 5), (London: MoD, 2004).

48 *Ibid.*, p. 13.

49 *Ibid.*

50 HM Government, *National Security Strategy and Strategic Defence and Security Review 2015: A Secure and Prosperous United Kingdom*, Cm 9161 (London: The Stationery Office, 2015).

51 The Covenant aims to address any perceived disadvantages faced by the armed forces community in comparison to other members of society.

Bibliography

Anderson, EW and Anderson, LD (2014), *An Atlas of Middle Eastern Affairs* (London: Routledge).

Bell, U (2000), *Sierra Leone: Building on a Hard-Won Peace* (UN Chronicle No. 4).

Bowcott, O (2012), 'Charles Taylor aided and abetted Sierra Leone war crimes, Hague court finds', *Guardian*, 26 April 2012.

Connaughton R (2001), 'Military Intervention and Peace-keeping: The Reality,' in *Joint Forces Quarterly Review*, 26 July 2001.

Chilcott, J (2016), 'The Report of the Iraq Inquiry', *House of Commons Report* HC264.

Cordesman, AH (2003) *The Iraq War: Strategy, Tactics, and Military Lessons* (Washington DC: CSIS).

Denov, M (2010), *Child Soldiers: Sierra Leone's Revolutionary United Front* (Cambridge: Cambridge University Press).

Farrell, T, Rynning, S, and Terriff, T (2013), *Transforming Military Power since the Cold War: Britain, France, and the United States, 1991–2012* (Cambridge: Cambridge University Press).

Fry, R and Bowen, D (2011), 'UK national Strategy and Helmand,' in Clarke (ed.), *The Afghan Papers: Committing Britain to War in Helmand, 2005–6* (London: RUSI).

Gberie, L (2005), *A Dirty War in West Africa: The RUF and the Destruction of Sierra Leone* (Indiana: Indiana University Press).

Government of Sierra Leone (1999), Lomé Peace Agreement. Freetown, Government of the Republic of Sierra Leone, www.sierra-leone.org/lomeaccord.html, accessed 10 August 2017.

HM Government (2015), *National Security Strategy and Strategic Defence and Security Review 2015: A Secure and Prosperous United Kingdom*, CM9161 (London: The Stationery Office).

International Justice Monitor, 'A project of the Open Society Justice Initiative', www.ijmonitor.org/charles-taylor-background/#fourteen, accessed 2 October 2017.

Keen, D (2005), *Conflict & Collusion in Sierra Leone* (Oxford: James Currey Publishers).

Kincaid, B (1999), *Dancing with the Dinosaur: How to do Business with the MoD in the Smart Procurement World* (Newcastle-upon-Tyne: UK Defence Forum).

Louth, J (2010), *A Low Dishonest Decade: Smart Acquisition and Defence Procurement into the New Millennium* (Cardiff: UWIC).

Louth, J (2016), 'Defence Acquisition Reform and the British Condition Promises, Betrayal and Resignation', in Kevin Burgess and Peter Antill (eds), *Emerging Strategies in Defense Acquisitions and Military Procurement* (Hershey, US: IGI Global).

Louth, J and Quentin, P (2014), 'Making the Whole Force Concept a Reality', *RUSI Briefing Paper* (London).

Louth, J and Taylor, T (2015), 'Beyond the Whole Force: The Concept of the Defence Extended Enterprise and its Implications for the Ministry of Defence', *RUSI Occasional Paper* (London).

Ministry of Defence (1998), *Strategic Defence Review: Modern Forces for the Modern World* (London: The Stationary Office).

Ministry of Defence (2002), *The Strategic Defence Review: A New Chapter, Vol. 1*, CM5566 (London: The Stationary Office). Can be accessed online at: http://archives.livreblancdefenseetsecurite.gouv.fr/2008/IMG/pdf/sdr_a_new_chapter_cm5566_vol.1.pdf, accessed 9 August 2017.

Ministry of Defence (2002), *The Smart Acquisition Handbook* (Edition 4), (London: MoD).

Ministry of Defence (2004), *The Smart Acquisition Handbook* (Edition 5), (London: MoD).

Ministry of Defence (2006), Joint Warfare Publication (JWP) 0-01, *British Defence Doctrine* (London: MoD).

Ministry of Defence (2011), 'The Armed Forces Covenant', May 2011.

NATO (1996), Glossary of Terms and Definitions, 1996.

Olonisakin, F (2015), 'United Nations Mission in Sierra Leone (UNAMSIL)', in Koops, Koops, MacQueen, Tardy, Williams (eds), *The Oxford Handbook of United Nations Peacekeeping Operations* (Cambridge: Cambridge University Press).

Pham, J (2005), *Child Soldiers, Adult Interests: The Global Dimensions of the Sierra Leonean Tragedy* (New York: Nova Publishers).

Prins, G (2002), *The Heart of War* (London: Routledge).

Richards, D (2011), 'Sierra Leone: Pregnant with Lessons', in Richards and Mills (eds), *Victory Among People: Lessons from Countering Insurgency and Stabilising Fragile States* (London: RUSI).

Schwartz, RA (2008), *Encyclopedia of the Persian Gulf War* (North Carolina: MacFarland).

Sheffy, Y (1993), 'The Military Dimension of the Gulf War', in Ami Ayalon (ed.), *Middle East Contemporary Survey*, Volume Xv: 1991 (Oxford: The Moshe Dayan Center).

Tucker-Jones, A (2014), *The Gulf War: Operation Desert Storm 1990–1991* (Barnsley: Pen and Sword).

Woods, LJ (2010), *Military Interventions in Sierra Leone: Lessons from a Failed State* (Kansas: DIANE).

United Nations Security Council, Resolution (UNSCR) 660, August 1990, https://documents-dds-ny.un.org/doc/RESOLUTION/GEN/NR0/575/10/IMG/NR057510.pdf?OpenElement, accessed 8 November 2017.

United Nations, Supplement to the London Gazette, 29 June 1991 www.thegazette.co.uk/London/issue/52589/supplement/38/data.pdf, accessed 8 November 2017.

2   Defence as policy and politics

Introduction

In any country, the basis of a relevant and efficient defence is a policy, which usually must be explicitly and carefully articulated and available to a wide number of stakeholders if it is to reach its full potential. How this policy is generated, its purposes, components, and constraints, is the subject of this chapter. For the authors, by its nature, it is a political endeavour.

The word 'policy' is far from standard or precise in its meaning, and in the realm of national politics it shares overlapping ground with the term 'strategy'. At heart policies are concerned with what an organisation is seeking to achieve, so in specific fields a government might be aiming at reducing national illiteracy rates or the numbers killed each year in road accidents. Policy in these terms can be articulated in very high level and indeed ambitious terms. The 2008 National Security Strategy (NSS) of the UK spelled out the government's 'single overarching national security objective of protecting the United Kingdom and its interests, enabling its people to go about their daily lives freely and with confidence, in a more secure, stable, just and prosperous world'.[1]

The 2015 National Security Strategy and Strategic Defence and Security Review (NSSSDSR) again began with the ambition and aims of the government:

> Our vision is for a secure and prosperous United Kingdom, with global reach and influence. Everything we do in the UK and around the world is driven by our determination to protect our people and our values, and ensure that our country prospers.[2]

However, persuasive and useful policy documents must deal not only with aspiration, but must also articulate the problems and challenges that need to be addressed, and how ambitions are to be achieved. It is by dealing with the 'how' that policy blends into strategy and political choice. To illustrate, two long-standing elements of UK defence policy stress that UK security requires a close and cooperative relationship with the United States and the deployment of a national nuclear force for deterrence purposes. These are presented as key elements regarding the implementation of defence.

Chapter objectives

By the end of this chapter the reader will understand:

1 The British approach to defence policy generation and the politics that constrain the choices made;
2 The emergence of a combined defence and security posture;
3 The broad purpose of defence and security policies;
4 Defence policy practice as historical constants and constraints.

Chapter structure

We start with a recent historical review of the British political class's approach to the development of defence policy. Thereafter, the narrative explores how emerging homeland security and resilience strategies have complicated this work. The authors locate this very modern debate within an understanding of the supposed purposes of defence policy making before considering the notion of defence policy as actually a set of political trade-offs between constant ambitions and imperatives and complicated and, occasionally, unwelcomed constraints.

The British approach: defence policy generation and review

The precise use of defence policy statements and reviews by the British government has not been fixed since the beginning of the Cold War, but has evolved as a practice of work and body of knowledge. Until 1998, the pattern was very much one of major and comprehensive defence 'reviews' being held at unequal and sometimes extended intervals. The timing of these was driven less by changing circumstances in the external world and more by recognition that existing British defence policies and practices had become unaffordable and that cuts in ambition had to be made. Defence reviews needed to be integrated particularly with the UK's gradual abandonment of the large empire it still possessed in 1945. Table 2.1 provides a summary of the five most prominent reviews before 1998 and the decisions associated with them.[3] Most involved commitments to reductions in the scope of defence activities but, crucially, there was also reference to the savings that were to be made through efficiencies generated by improved organisation and management.

Put simply, the major defence reviews recognised that the former policy and its commitments had become unaffordable, but asserted that its replacement would be affordable because of the savings that the Ministry of Defence would create. This pattern of argument was to be continued after the end of the Cold War.

However, policy and its application did not remain constant in the intervening periods between the reviews. Specifically, as the Cold War evolved, each year the MoD produced a Statement on the Defence Estimates, often with a second

Table 2.1 Summary of pre-1998 defence reviews

Date	Defence Minister	Chief findings
1957	Duncan Sandys	Defence against and deterrence of the Soviet Union were defined as the central problem. While NATO and US ties were at the centre of UK policy, there was increased reliance on missiles and the nuclear dimension of deterrence. The end of conscription was announced, meaning cuts to the Army, which was reduced to 160,000.[4]
1966	Denis Healey	Defence spending was to be frozen and the UK was to scrap an aircraft carrier. UK would not maintain foreign bases against the wishes of local governments and the Aden base was to be given up. There was stress on the UK's peacekeeping role in the Middle East and Asia.[5]
1968	Denis Healey	UK was to abandon aircraft carriers, and give up its military responsibilities east of the Suez Canal, leading to the independence of the Gulf Sheikhdoms and the closure of UK military facilities in Singapore.[6]
1974–5	Roy Mason	There was further focus on the challenges of dealing with the Warsaw Pact and four UK commitments were deemed essential: the contribution to frontline forces in Germany; anti-submarine activities in the North Atlantic; the UK's nuclear deterrent; and home defence. Britain's out of area capability to be further reduced.[7]
1981	John Nott	Significant reductions in the size of the surface fleet and emphasis on the UK's continental commitment to NATO.[8]

volume containing statistical information. These were used to announce tweaks in the different aspects of defence policy, concerning such issues as the perceived evolution of the Soviet threat, the importance to be ascribed to cooperation with European states, and the (diminishing) resources that could be allocated for 'out-of-area' missions.

In terms of political cohesion, not least in the Conservative Party, the 1981 review proved very divisive and painful. The minister's conclusion that the size of the Navy in terms of ships and people should be significantly reduced was opposed by many Conservatives. The Navy Minister of the time, (now Sir) Keith Speed refused to withdraw his opposition to the proposed cuts, which led to his dismissal by Prime Minister Thatcher.[9] Soon after, organisational change ended the appointment of junior ministers for the individual services. Given that the Argentinian invasion of the Falklands soon provided a central stage for naval capabilities, the overall result of the 1981 process and subsequent events could have induced a reluctance by the Thatcher and then Major governments to undertake any further major defence reviews. Even the end of the Cold War did not stimulate such an exercise, with the government restricting itself to an examination of how money could best be saved in defence without compromising visible capabilities. The resulting documents were *Options for Change* in 1990[10] and *Front Line First: the Defence Costs Study* in July 1994.[11]

In the 1997 election campaign, the opposition Labour Party taunted the Conservatives about their failure formally to adjust defence policy by reviewing defence in the new world. When Labour won the election, the Blair government had to conduct such a review itself, a process that took a little over a year and led to the publication of the Strategic Defence Review (SDR) in July 1998.[12] The core document covered 56 pages and in addition there were 11 Supporting Essays.

In terms of scope, content, and process, the SDR set a new standard for defence reviews and defence policy, and it aroused much interest from many other governments. In terms of specifying the UK's role in the world and the missions of its armed forces, it was of lasting impact since subsequent reviews did not significantly modify the content of specified UK policy. Only the scale of ambition became somewhat muted.

The Labour government maintained a stream of modifications and updates to the SDR with a 57 page Defence White Paper in 1999;[13] a brief Defence Policy statement in 2001;[14] a document following the September 2001 attack on the World Trade Center entitled 'The Strategic Defence Review: A New Chapter' in July 2002, which addressed SDR implementation and the particular challenges of dealing with terrorism;[15] and then a further two volume White Paper in December 2003 called 'Delivering Security in a Changing World'.[16] More flesh and detail was put on this a year later with 'Delivering Security in a Changing World: Future Capabilities'.[17] However, none of these involved the range of consultation and scrutiny of fundamentals that had been a feature of the 1997–1998 exercise and, despite the move to locating defence explicitly within the context of a wider National Security Strategy in 2008,[18] in the 2010 election the Conservatives followed the Labour example of more than a decade earlier and criticised the government for not having fully reviewed defence policy since 1998. Once in power, the Conservative-Liberal coalition had to live up to its commitment to what it was to specify as the Strategic Defence and Security Review (SDSR).

The major thrust of Conservative criticism of Labour's treatment of defence was that it had allowed a large gap (a black hole) to appear between planned defence commitments and the resources likely to be available for defence (especially after the 2007–2009 financial crisis had disrupted government spending plans). The Conservative-Liberal Democrat coalition therefore felt it had to generate a defence policy compatible with the public finance spending plans (the Comprehensive Spending Review)[19] due in the autumn of 2010. Elected in April 2010, the coalition thus had only a few months to produce a new policy statement that located defence within the wider security space.

The consequence was that the defence review promised by the coalition virtually took place in three stages of which the SDSR document in 2010 was only the first. The second was the Levene Report in 2011 which laid out how the organisation and management of the MoD was to be changed to deliver improved performance.[20] The headlines of this report included the commitment to reduce the size of the central Ministry and to allocate more financial power and responsibility to the single services. In particular, they were to become financially responsible for the support of their own equipment. The third stage was the publication of the National Security Through Technology White Paper in 2012[21] that was intended to strengthen defence acquisition practices and to take over from the Defence Industrial Policy document that the Labour government had introduced in 2005. The National Security through Technology paper publicly strengthened the government's ambition to buy defence equipment on a competitive basis from the global market, although it did acknowledge that security considerations might sometimes make this inappropriate.[22] The relationships among these reports again involved emphasis that the UK's defence ambitions and activities were to be rendered affordable through improvements in efficiency. Thus, 'the role of defence management improvements was not just to make better use of public money, but to enable the government to avoid a significantly reduced UK military role in the world'.[23]

This reliance on improved efficiency to render affordable defence policy was a feature also of the awkwardly titled National Security Strategy and Strategic Defence and Security Review (NSSSDSR) of 2015[24] following the election of a one-party Conservative government in that year. Under the Comprehensive Spending Review settlement covering the years out to 2020–2021, the MoD was to make:

> savings of £9.2 billion, including £2 billion from pay restraint, £2.1 billion from improved commercial terms in the Equipment Plan, and reprioritisation of £2 billion of existing funding, all of which will be reinvested to fund the SDSR commitments.[25]

From 2010 it was also agreed that defence reviews should be conducted every five years, which translated into after every general election if a government served a full-term. This was not quite the same as finding that such reviews should be held whenever the world had changed sufficiently to justify one, but it did mean that government could not as easily avoid facing up to some difficult developments (like the end of the Cold War) and their defence implications. In between major reviews, the government reported on their implementation and could make some adjustments through the publication and content of the MOD's Annual Report and Accounts[26] (Ministry of Defence) and, from December 2016, the publication of an annual report on SDSR implementation.[27] The Statements on the Defence Estimates (SDE) have disappeared and the publicly available Defence Statistics that used to accompany the SDEs are now constantly updated and made publicly available only online.

From defence policy to defence within national security policy

The observant reader will have already noted that the terms 'security' and 'defence' have been used without any discussion of their relationship. There is an extensive literature on this subject and, while this book is focused on the defence sector, some words are needed.

During the Cold War, the major danger to UK and indeed Western societies appeared to come from the military challenges presented by the Soviet Union and its allies. The prime response to these military challenges was the generation of UK military forces which, linked to those of allies, would deter any possible attack. Even then, however, defence capabilities were often supplemented by diplomatic activities to improve East-West relations, not least in the field of arms control and disarmament. In organisational terms, the British MoD and Foreign and Commonwealth Office (FCO) worked together, with the MoD leading on conventional arms control and the FCO on nuclear issues. Thus, defence was never seen as entirely separate from the wider security picture.

In addition, even during the Cold War there were the first indications that at least some saw that the major challenges for the well-being of Western societies were not all military. In West Germany, for instance, the rise of the Green Party in the 1980s emphasised environmental threats to the well-being of Germans. In the UK, the problems with Northern Ireland after 1969 saw the need for the armed forces to act over the long-term in support of the police in the context of wider efforts to contain and even resolve the challenges presented by the divisions in the province. Thus 'security' came to include domestic threats of organised violence and serious environmental problems that raised issues of pollution and climate change.

Then, after the end of the Cold War, there was an increasing trend to view defence as a sub-element of the wider security picture and its problems. Thus, while the Labour government of 1997 initiated an explicit 'defence review', which led to the publication of the Strategic Defence Review (the SDR), in 2008 the Labour government published a National Security Strategy (NSS), which was followed by the coalition government's Strategic Defence and Security Review (SDSR) in 2010 and the Conservative's National Security Strategy and Strategic Defence & Security review (NSSSDSR) in 2015.

The concept of a 'security' problem in liberal democratic states has been widened, arguably by a mix of empirical developments in the wider world and by political considerations. Problems that do not involve malice as such, such as climate change and infectious diseases, have come to prominence as things that could have a significant negative effect on many members of society. They have thus been placed within the security area, in contrast for instance to changing patterns of drug use in the UK.

But it should also be noted that 'security' problems are almost by definition significant and therefore should receive attention and resources. That means that those concerned with a particular issue can be motivated to get it placed in the security sphere. If they can have their focus of interest 'securitised', increased funding is likely. Thus, special interest groups broadly welcome their inclusion on the security agenda.

In practical terms, the list of 'security challenges' identified in the 2008 NSS comprised the following:

- Terrorism;
- Nuclear weapons and other weapons of mass destruction;
- Transnational organised crime;
- Global instability and conflict, and failed and fragile states;
- Civil emergencies;
- State-led threats to the UK.[28]

The latter were dealt with summarily:

> Our assessment remains the same as in the 1998 Strategic Defence Review: for the foreseeable future, no state or alliance will have both the intent and capability to threaten the United Kingdom militarily, either with nuclear weapons or other weapons of mass destruction, or with conventional forces.[29]

Needless to say, any efforts to deal with these challenges had to take account of their perceived causes, and the drivers of security problems were classified under:

- Challenges to the rules-based international order;
- Climate change;
- Competition for energy;
- Poverty, inequality, and poor governance;
- Global trends (including some effects of globalisation).[30]

The 2010 and subsequent national security reviews presented a more detailed list of threats and problems, categorised into three tiers relating to state-on-state conflict, terrorism, and military aid to the civil authority to assist with disaster relief.

Therefore, there appears a need to record that security and defence reviews should, and in the UK do, address two dimensions:

> What are the military threats and challenges that a country faces, and how they are to be dealt with in terms of both military and non-military techniques?

> What are the non-military security threats and serious challenges that country faces, and what is to be the role of the armed forces and other resources of the MoD in efforts to deal with them?

The purposes of defence policies

For the authors, there are a number of identifiable objectives and purposes for defence policy.

A means of political direction of the armed forces

In principle, a country's defence policy may be neither in the public domain or even written down. It can take the form of a series of implicit assumptions and beliefs that form a significant element of the culture of a government's defence sector. However, a written policy available to the public and indeed to overseas audiences, as well as to all elements of government, can serve four valuable purposes: to shape the development of the armed forces; to promote the civilian and political direction of defence; to inform national stakeholders and publics; and, so, send messages to international audiences. Documented British defence policies have filled all these roles.

British defence policies and their generation have supported civilian and political direction of the military instead of simply allowing the different sections of the armed forces to build the capabilities and systems that they themselves appreciate. While the uniformed military have expertise in the conduct of military operations and the possible capabilities of potential adversaries, they do not have a monopoly on, or even primary expertise on, the intentions of foreign governments and groups, on the non-military consequences of military actions and stances, on the opportunity costs for the economy and wider society associated with military spending, or on how the military instrument should be best integrated with other tools of state influence such as economic aid and trade rights. They are often not very good at deciding what would be the best equipment for their units.

In the UK, the processes of generating defence policy (in which the military voice still needs clearly to be an important contributor) have and continue to support political and civilian direction and control of the military. The very important 1998 SDR was the product to some extent of a national debate involving a wide range of interested parties.[31] The move to locating defence policy within wider national security strategy has required the involvement of many civilian parts of government, indeed the 2015 NSSSDSR was Cabinet Office led. Finally, within the MoD itself, an important contribution of civil servants is to assist ministers in the generation and adjustment of policy matters.

Communicating with national stakeholders

Second an explicit defence policy should explain to the legislature, the interested public, and the media how the taxpayer's money is to be used in defence and for what reasons. In the UK, the legislature has long approved the budget proposed by the government, but there is no automatic assurance that this will always be the case. In other countries, the situation is very different, with different elements in the US Congress sometimes having disparate views about what the defence budget should be and how it should be spent. If parliamentarians and voters are to be active and committed in their roles as representatives of citizens and taxpayers, they are entitled to be informed about the use of the money they hand over and a country's stated defence policy is an important document in this regard. Citizens kept in the dark about the use of their taxes with regard to defence are more likely to resent the payments they make. Moreover, governments often want active popular support from their populations for the armed forces, if only to encourage talented young people to offer themselves as employees to a military that they can see is clearly relevant and important to national problems. A reasoned defence policy statement is a key element in the public explanation of and justification for defence activities.

A significant feature of British defence policies has been their public and detailed nature. Moreover, the arrival of the internet has removed the dependence of the interested public, including academic researchers, on their ability to secure printed copies.

In the UK, among other investigations, the Defence Committee of the House of Commons investigates and reports on all the major policy products of the MoD.[32] The National Audit Office frequently supports its work with its own studies.

Communicating with foreign governments and audiences

Next, a defence policy document is a source of communication with external players, which include allies, current or potential adversaries, defence suppliers, defence customers, and international organisations such as NATO and the United Nations. As well as offering facts and statements about aspirations, a defence policy statement can be used to provide clarity and remove ambiguity, to offer re-assurance, to reinforce deterrence relationships, and even in some circumstances to support intimidation.

Anyone reading British defence reviews and statements may well be struck by their emphasis on the capabilities and significance of the UK, reminding the reader that the UK had the fifth biggest economy in the world[33] and had the fifth largest defence budget,[34] messages clearly not meant solely for audiences within the British government. Chapter 2 of the 2015 NSSSDSR proclaimed the UK as 'Strong, Influential, Global'[35] and went on to assert the respect and impact of numerous UK institutions and organisations, inside the security domain and in the broader soft power domain. Seventeen years earlier Secretary of Defence George Robertson had told the nation and the world:

> The British are, by instinct, an internationalist people. We believe that, as well as defending our rights, we should discharge our responsibilities in the world. We do not want to stand idly by and watch humanitarian disasters or the aggression of dictators go unchecked. We want to give a lead, we want to be a force for good.[36]

Shaping defence capabilities and the armed forces

Finally, and probably most significantly, in addition to supporting the civilian direction of defence and sending important messages to the nation and the world, defence policy statements, by spelling out challenges, ambitions, and aspirations and, at least in broad terms, how they are to be addressed, guide how the armed forces should set their priorities and the sorts of capabilities they should develop. Clear defence policies also enhance the chances for coherent and coordinated action on the part of the different branches of the armed forces.

In the 1980s and even into the 1990s an important policy choice was that UK resources should be increasingly focused on the British contributions to the defence of NATO territory. Even the 1991 Statement on the Defence Estimates[37] included the following conclusions:

> The prime focus of our defence effort must continue to be the collective security provided by the North Atlantic Alliance....[38]
>
> We judge that our armed forces should retain their current roles in the alliance: the provision of nuclear forces; direct defence of the United Kingdom; a contribution to the defence of the European mainland; and maritime defence of the eastern Atlantic and Channel.[39]

As far as 'contingencies outside the NATO area' were concerned, UK capabilities were to continue to be just a consequence of 'the inherent flexibility and mobility of our future force contribution to NATO'.

Only the 1998 Strategic Defence Review spelled out for the first time that UK forces should be committed largely to force projection in support of international security and that defence of the NATO area against state threats was not a pressing need. The 1998 policy provided the rationale for the acquisition of many hardware assets, including more large transport aircraft, improved global satellite communication, lighter armoured vehicles and, of course, two aircraft carriers. Chapter 3 and Supporting Essay 6 of the SDR summarised eight missions for UK forces (support for peacetime security, the protection of British Overseas Territories, implementing Defence Diplomacy, supporting wider British interests, conducting peace support and humanitarian operations, dealing with regional conflict outside the NATO area, responding to regional conflict inside the NATO area, and deterring or dealing with a strategic attack on NATO). The specification of these missions enabled the articulation of 28 'Military Tasks' and the conception of the capabilities needed for their execution.

While the UK's scale of ambition has been moderated, the broad direction and scope of UK defence desired capabilities as laid out in the SDR have not been significantly altered by subsequent reviews. The SDR included a short paragraph that could have been inserted into subsequent policy statements:

> We could of course, as a country, choose to take a narrow view of our role and responsibilities which did not require a significant military capability. This would mean that we would not wish and would not be able to contribute effectively to resolving crises such as Bosnia, Kosovo, or invasion of Kuwait. This is indeed a real choice, but not one the Government could recommend for Britain.[40]

While the 1998 review signalled some significant changes of direction for UK defence policy, it also left in place two constants, already noted. The first was that the UK should be an 'independent' nuclear power and the second was the emphasis on the United States as the most important, indispensable ally of the UK.

Policy constants

The UK as a nuclear power

The UK, having made a significant input into the development of nuclear weapons in the United States, pressed ahead with the development of a national nuclear weapon in the 1940s and 1950s in the face of initial American opposition. Having developed three separate bombers to deliver its weapons (Valiant, Victor and Vulcan), Britain then settled under the Bermuda agreement with the United States in 1959 on the purchase of Polaris missiles equipped with British warheads and their deployment on British-built nuclear submarines. When Polaris had to be replaced, the 1980–1981 decision was to continue the deterrent through the leasing of Trident missiles, the building of larger Vanguard class submarines, and again the fitting of those missiles with UK warheads. In both 2006 and 2016, parliament approved the development and production of a second replacement submarine, the Dreadnought class, to take upgraded Trident missiles.

In policy terms, the clear role of these nuclear forces during the Cold War was to deter a nuclear attack on the UK. In addition, however, the UK's deterrent provided a second centre of decision-making (and thus enhanced credibility) for the NATO commitment under its Flexible Response doctrine to be ready to escalate a conventional war to prevent a Warsaw Pact takeover of Western Europe. After the end of the Cold War, the deterrence of a nuclear attack on the UK remained an agenda item, and the ISS SDSR of 2015 maintained the long-standing policy emphasis.

> The UK's independent nuclear deterrent will remain essential to our security today, and for as long as the global security situation demands. It has existed for over 60 years to deter the most extreme threats to our national security and way of life, helping to guarantee our security, and that of our allies.[41]
>
> Other states continue to have nuclear arsenals and there is a continuing risk of further proliferation of nuclear weapons. There is a risk that states might use their nuclear capability to threaten us, try to constrain our decision making in a crisis or sponsor nuclear terrorism. Recent changes in the international security context remind us that we cannot relax our guard. We cannot rule out further shifts which would put us, or our NATO Allies, under grave threat.[42]

The policy arguments used here could of course be adopted by almost any state, including the Netherlands or Germany. The change of direction of Russian policy under President Putin has given them a weight that they lacked between 1990 and 2014, but the broader reality is that the British nuclear treadmill is one from which it is difficult to step off. If the UK abandoned its nuclear weapons and the world did change significantly for the worse, it would take a long time and a lot of money to reconstitute the capability, and a decision to reconstitute would likely be very difficult in national and international political terms. On the other hand, continuing with the deterrent appears to cause little international or even national fuss, even taking into account the activities of the Campaign for Nuclear Disarmament. However, a continually-at-sea submarine-based system has presented a significant financial bill and the Successor programme is a major financial challenge to the government.[43]

The United States as the essential ally

The second point of policy constancy before and after the Cold War has been that the United States is the UK's most important ally, indeed US support has been seen as a requirement for UK security: 'The partnership with the United States is our most important bilateral relationship and central to our national security, including through its engagement with NATO'.[44] The reliance on the United States has often been couched largely in terms of enthusiasm for NATO and indeed this was a feature of the 1998 Strategic Defence Review.

Before and after the Cold War, the UK has seen it as essential to sustain American commitment to Europe's defence and has also hoped-for influence in Washington by being the ally most valued by the United States. British policy has been to seek influence with American administrations by being able to make a significant contribution to any US-led operation, a stance that was apparent in the division-sized land force involvement in the 1991 campaign to liberate Kuwait.

The SDR Supporting Essay 3 on Technology included words that had important procurement and industrial implications. Discussing the developments associated with the Revolution in Military Affairs, the Essay observed:

> It is clear that exploiting these technologies will lead to significant improvements in military capability. These will inevitably be led by the US. If Britain and other allies can successfully tap into these developments, the result will be more effective coalition operations. Conversely there is potential for multinational operations to become more difficult if compatible capabilities are not preserved. This could lead to political as well as military problems. *Our priority must therefore be to ensure that we maintain the ability to make a high-quality contribution to multinational operations and to operate closely with US forces throughout the spectrum of conventional operations.* To do this we may need to be selective about the technologies we develop nationally or on a European basis, and be prepared to use US technologies in other areas in order to make a leading contribution to multinational operations.
>
> (author's emphasis)[45]

Concern not to weaken the US commitment to Europe has led the UK to approach cooperation with Europeans, whether on a bilateral, multilateral, or institutionalised basis, with some wariness and, in policy statements, cooperation with the United States has almost invariably been dealt with before the value of working with Europeans is assessed.[46]

The British psychological and material dependence on the United States clearly carries risk. The Obama administrations spoke of a 'Pivot to Asia'.[47] While the consequences of that announcement were much affected by changed Russian behaviour, including its annexation of the Crimea, the 2016 election of a president from outside the mainstream of US politics signalled that there are forces at work in the United States that might radically change the US role in the world.

The reducing likelihood of unilateral UK operations

Even before the end of the Cold War, the increasing policy assumption has been that UK forces are unlikely to be engaged in operations on a national basis: multinational operations will be the rule, either in a NATO, European, or ad hoc framework. This has meant that the UK has acknowledged that it need not seek the full spectrum of capabilities and has recognised that in some fields it must rely on the United States in particular.

> The full spectrum of capabilities is not required for large scale operations, as the most demanding operations could only conceivably be undertaken alongside the US, either as a NATO operation or a US-led coalition....
>
> (Ministry of Defence July 2004, p. 2)[48]

Clearly this argument reinforces the apparent viability of British readiness to rely on the United States, while resting on the assumption that there is no significant threat to any UK overseas territories, most obviously the Falkland Islands. However, the expectation of multilateral operations only has not led to any national role specialisation arrangements with European allies.

Constraints on the armed forces

In defence policy, national aims and how they are to be achieved must also make reference to the constraints and direction that the government wishes to apply to the armed forces, in matters as varied as the representation of members of ethnic minorities in the uniformed services, the rights of women to serve today in all branches of the armed forces, and the desirability of a proportion of defence contracts going to small and medium-sized enterprises. Defence policy can thus address specific and detailed topics, including the sorts of weapons that the armed forces are not allowed to acquire and use: it is UK policy that UK forces should not use anti-personnel mines or cluster weapons and, consequently, the UK is a party to the 1997 Ottawa Treaty, the Anti-Personnel Mine Ban Convention, whereas its key ally, the United States, is not.

So, all aspects of national 'defence policy' may not necessarily be captured in a formal defence policy statement or review. Some issues may justify their own documented specification and justification, and some may even require incorporation into legislation (most obviously the Armed Forces Act) before they can come into effect. As an example, the British prime ministerial decision that women should be able to serve in all army units, including the infantry, a clear defence policy matter, was taken outside the SDSR framework in July 2016.

Ambiguity and what lies beneath

Obviously, the wording of a document that is meant to provide messages to people inside national defence and within the national security community within a country, to parliament and the public, and to a range of foreign audiences, needs to be drafted with great care. It is not surprising that ambiguity and even contradiction can be found within policy statements. To support the implementation of publicly-available policy, contemporary British governments have generated, beneath the public statement, a classified piece of work, currently designated as Defence Strategic Direction, which provides more detailed guidance on defence techniques and priorities for use within government. Another British classified supplement to the public face of defence policy is the collection of a significant number of centrally approved, credible scenarios for military operations that could take place given the state of the world and British defence policy. These scenarios can then be used to assess the utility of candidate future investments, for instance in different types of equipment.

Conclusion

This chapter has examined British defence policy through scrutiny of the idea and purposes of policy through the practice of periodic major reviews and annual policy reports and adjustments, and through the identification of constant and changing elements. As argued, despite the multiplicity of defence white papers and policies noted here, there have been themes of considerable consistency.

During the Cold War the message was that the UK needed to focus increasingly on its NATO commitments on the inner German frontier, on the North Atlantic, and on its nuclear deterrence. With reluctance, the commitments left from the days of empire were repeatedly scaled back. Once the Cold War ended, the central theme was that the UK would be a major defence contributor to security in Europe and beyond. The 1998 SDR sense that the UK would not act 'east of Suez' or at least east of the Gulf was abandoned as the UK intervened in Afghanistan after the Cold War.

Moreover, successive British governments have rarely been tempted to play down the significance of the UK or the reduced resources available compared with other states. UK policies have been conspicuously ambitious, although commitments to empire had largely to be abandoned. In 2018 the UK still aims to be both a nuclear power with a submarine-based deterrent and a state able to project and sustain sizeable conventional forces around the world for a range of missions.

One strategic criticism that might be levelled at UK defence policy is that it has been based increasingly since the Second World War on the assumption that the United States will always be there to support the UK.

In 2015, the NSSSDSR reiterated:

> Our special relationship with the US remains essential to our national security. It is founded on shared values, and our exceptionally close defence, diplomatic, security and intelligence cooperation.[49]

Note the word 'essential': no contingency plans have been developed to deal with the possibility of a change of direction in the United States and, as a consequence, the number one fear of many political figures and officials has often been that of alienating the United States. When parliament voted against intervention in Syria in 2013, the reaction of then Foreign Secretary Philip Hammond was concern about what the choice would do to the 'special relationship' with the United States.[50] Successive British governments have been content to rely heavily on the United States for military security while, at the same time, apparently fearing that the US commitment might not be robust.

The ambition associated with UK declaratory policy has also consistently been accompanied by aspiration to make the defence sector more efficient. During the Cold War, major reviews were triggered by financial commitments much outstripping the likely funds available, but governments remained keen to minimise the reduction in the UK's role in NATO and the world. Although changed world circumstances justified the SDR in 1998, the MoD even then was looking at a 'bow wave' of growing commitments in the equipment plan. A significant part of the response was to improve defence management. In 2010 the Conservative-Liberal Democrat coalition was daunted by the £38 billion black hole that it found in the equipment plan: some clear cuts and cancellations were made (notably to the Harrier and Nimrod fleets) but, again, there was an intention to improve the running of the sector. Five years later, the commitments in the 2015 NSSSDSR were presented as practicable for an MoD being allocated 2 per cent of GDP because, over a five-year period, the MoD committed itself to making £9.1 billion in efficiency savings. Thus, defence policy was crucially linked to improved defence management, and it is to this topic that attention is turned next.

Notes

1 Cabinet Office, *National Security Strategy of the United Kingdom: Security in an Interdependent World*, (London: The Stationary Office, 2008), p. 5.
2 HM Government, *National Security Strategy and Strategic Defence and Security Review 2015: A Secure and Prosperous United Kingdom*, CM9161 (London: The Stationery Office, 2015), para 1.1, p. 9.
3 See, for instance, Claire Taylor, *A Brief Guide to Previous UK Defence Reviews* (London: House of Commons Library, October 2010).

4 Ministry of Defence, *Defence White Paper: Outline of Future Policy*, CM124 (London: The Stationary Office, 1957).

5 House of Commons Command Paper, [2901], *Statement on the Defence Estimates 1966, Part I: The Defence Review* (February 1966); CM2902, *Statement on the Defence Estimates 1966, Part II: Defence Estimates 1966–67* (February 1966).

6 House of Commons Command Paper, [3701], *Supplementary Statement on Defence Policy 1968*.

7 1974–1975 Defence Reviews are often referred to as the Mason Review.

8 House of Commons Command Paper, [8288], *The United Kingdom Defence Programme: The Way Forward*, 1981.

9 Trevor Taylor, 'Jointery and the Defence Review', in Michael Codner and Michael Clarke (eds), *A Question of Security: The British Defence Review in an Age of Austerity* (London: IB Tauris, 2011), pp. 175–86.

10 Ministry of Defence, *Options for Change: The UK Defence Review, 1990–1991* (London: The Stationary Office, 1990).

11 Ministry of Defence, *Front Line First: The Defence Cost Study*, (London: The Stationary Office, 1994).

12 Ministry of Defence, *Strategic Defence Review: Modern Forces for the Modern World* (London: The Stationary Office, 1998).

13 Ministry of Defence, *Defence White Paper 1999* (London: The Stationary Office, 1999).

14 Ministry of Defence, *Defence Policy 2001*, (London: The Stationary Office, 2001).

15 Ministry of Defence, *The Strategic Defence Review: A New Chapter, Vol. 1*, CM5566 (London: The Stationary Office 2002). The document was in two volumes, with the second providing supporting essays.

16 Ministry of Defence, *Delivering Security in a Changing World: Defence White Paper 2003*, CM6041-1 (London: The Stationary Office, 2003).

17 Ministry of Defence, *Delivering Security in a Changing World: Future Capabilities*, CM6269 (London: The Stationary Office, 2004).

18 Cabinet Office, *National Security Strategy of the United Kingdom: Security in an Interdependent World* (London: The Stationary Office, 2008).

19 HM Treasury, *Comprehensive Spending Review 2010* (London: HM Treasury).

20 Ministry of Defence, *Defence Reform: An independent report into the structure and management of the Ministry of Defence* (London: The Stationary Office, 2011).

21 Ministry of Defence, *National Security Through Technology: Technology, Equipment, and Support for UK Defence and Security*, CM8278 (London: The Stationery Office, 2012).

22 *Ibid.*, p. 8.

23 Taylor (2011), op. cit., p. 225.

24 See Chapter 7 of HM Government (2015), CM9161.

25 HM Treasury (2015), *Spending Review and Autumn Statement 2015*, CM 9162 (London: HM Treasury), www.gov.uk/government/news/ministry-of-defences-settlement-at-the-spending-review-2015, accessed 9 November 2017.

26 An index for MoD Annual Reports and Accounts is available at: www.gov.uk/government/collections/mod-annual-reports, accessed 4 November 2017.

27 Ministry of Defence, *Annual Report and Accounts 2015 to 2016* (London, 2016), www.gov.uk/government/publications/ministry-of-defence-annual-report-and-accounts-2015-to-2016, accessed 10 November 2017.

28 Cabinet Office (2008), op. cit., p. 10–16.

29 *Ibid.*, p. 15.

30 *Ibid.*, pp. 16–22.

31 Ministry of Defence (1998), op. cit.

32 House of Commons Defence Committee, *Flexible Response? An SDSR checklist of potential threats and vulnerabilities: Government Response to the Committee's First Report of Session 2015–16* (London: The Stationary Office, 2016).

33 HM Government (2015), CM9161, p. 5.
34 Ministry of Defence, *Annual Report and Accounts 2016/2017* (London: The Stationery Office, 2017), p. 21.
35 HM Government (2015), CM9161, p. 13.
36 Ministry of Defence (1998), op. cit., p. 4.
37 Ministry of Defence, *Statement on the Defence Estimates: Britain's Defence for the 1990s, Vol. 1*, CM1559-1 (London: The Stationary Office, 1991).
38 *Ibid.*, 40.
39 *Ibid.*, 47.
40 Ministry of Defence (1998), op. cit., p. 16.
41 HM Government (2015), CM9161, p. 34.
42 *Ibid.*
43 Ministry of Defence, *Dreadnought submarine programme: Factsheet*, 2016, www. gov.uk/government/publications/successor-submarine-programme-factsheet/successor-submarine-programme-factsheet, accessed 12 November 2017.
44 Cabinet Office (2008), op. cit., p. 8.
45 Ministry of Defence, *Strategic Defence Review*: Supporting essay 3, the impact of technology (London: The Stationary Office, 1998), para. 10.
46 Ministry of Defence, *The Future Strategic Context for Defence* (London: The Stationary Office, 2001), p. 13.
47 Mike Green, The Legacy of Obama's "Pivot" to Asia', *Foreign Affairs*, 3 September 2016.
48 Ministry of Defence, *Delivering Security in a Changing World: Future Capabilities*, CM6269 (London: The Stationary Office, 2004), p. 2.
49 HM Government (2015), CM9161, p. 14.
50 BBC News, 'Syria crisis: Cameron loses Commons vote on Syria action', 30 August 2013, www.bbc.co.uk/news/uk-politics-23892783, accessed 8 October 2017.

Bibliography

BBC News (2017), 'Syria crisis: Cameron loses Commons vote on Syria action', 30 August 2013, www.bbc.co.uk/news/uk-politics-23892783, accessed 8 October 2017.
Cabinet Office (2008), *National Security Strategy of the United Kingdom: Security in an Interdependent World* (London: The Stationary Office).
Green, M (2016), 'The Legacy of Obama's "Pivot" to Asia', *Foreign Affairs*, 3 September 2016.
HM Government (2015), *National Security Strategy and Strategic Defence and Security Review 2015: A Secure and Prosperous United Kingdom*, CM9161 (London: The Stationery Office).
HM Treasury (2010), *Comprehensive Spending Review* 2010 (London: HM Treasury).
HM Treasury (2015), *Spending Review and Autumn Statement 2015*, CM9162 (London: HM Treasury), www.gov.uk/government/news/ministry-of-defences-settlement-at-the-spending-review-2015, accessed 9 November 2017.
House of Commons Command Paper (February 1966), [2901], *Statement on the Defence Estimates 1966, Part I: The Defence Review*.
House of Commons Command Paper (February 1966), [2902], *Statement on the Defence Estimates 1966, Part II: Defence Estimates 1966–67*.
House of Commons Command Paper (1968), [3701], *Supplementary Statement on Defence Policy 1968*.
House of Commons Command Paper (1981), [8288], *The United Kingdom Defence Programme: The Way Forward*.

House of Commons Defence Committee (2016), *Flexible Response? An SDSR checklist of potential threats and vulnerabilities: Government Response to the Committee's First Report of Session 2015–16* (London: The Stationary Office), www.publications. parliament.uk/pa/cm201516/cmselect/cmdfence/794/794.pdf, accessed 4 October 2017.

Ministry of Defence (1957), *Defence White Paper: Outline of Future Policy*, CM124 (London: The Stationary Office).

Ministry of Defence (1990), *Options for Change: The UK Defence Review, 1990–1991* (London: The Stationary Office).

Ministry of Defence (1991), *Statement on the Defence Estimates: Britain's Defence for the 1990s, Vol. 1*, CM1559-1 (London: The Stationary Office).

Ministry of Defence (1994), *Front Line First: The Defence Cost Study*, (London: The Stationary Office).

Ministry of Defence (1998), *Strategic Defence Review: Modern Forces for the Modern World* (London: The Stationary Office).

Ministry of Defence (1999), *Defence White Paper 1999* (London: The Stationary Office).

Ministry of Defence (2001), *Defence Policy 2001* (London: The Stationary Office).

Ministry of Defence (2001), *The Future Strategic Context for Defence* (London: The Stationary Office).

Ministry of Defence (2002), *The Strategic Defence Review: A New Chapter, Vol. 1*, CM5566 (London: The Stationary Office).

Ministry of Defence (2003), *Delivering Security in a Changing World: Defence White Paper 2003*, CM6041-1 (London: The Stationary Office).

Ministry of Defence (2004), *Delivering Security in a Changing World: Future Capabilities*, CM6269 (London: The Stationary Office).

Ministry of Defence (2011), *Defence Reform: An independent report into the structure and management of the Ministry of Defence* (London: The Stationary Office).

Ministry of Defence (2012), *National Security Through Technology: Technology, Equipment, and Support for UK Defence and Security*, CM8278 (London: The Stationery Office).

Ministry of Defence (2016), *Dreadnought submarine programme: Factsheet*, www.gov. uk/government/publications/successor-submarine-programme-factsheet/successor-submarine-programme-factsheet, accessed 12 November 2017.

Ministry of Defence (2016), *Annual Report and Accounts 2015 to 2016* (London: The Stationery Office).

Ministry of Defence (2017), *Annual Report and Accounts 2016/2017* (London: The Stationery Office).

Taylor, C (2010), *A Brief Guide to Previous UK Defence Reviews* (London: House of Commons Library, October 2010), researchbriefings.files.parliament.uk/documents/ SN05714/SN05714.pdf.

Taylor, T (2011), 'Jointery and the Defence Review', in Michael Codner and Michael Clarke (eds), *A Question of Security: The British Defence Review in an Age of Austerity* (London: IB Tauris), pp. 175–86.

3 Defence as management

Introduction: the reluctant manager

Military officers in the UK, young and senior alike, often do not like to think of themselves as 'managers', preferring the role of 'leader', an inclination reinforced by the stress in the officer selection process on 'leadership potential'.[1] Perhaps only when officers are about to leave the armed forces and hoping to secure new employment might they be inclined to deconstruct their careers to underline the management experience they have acquired in uniform.

Yet the efficient, effective, and successful preparation and use of armed forces has always rested centrally on what we would today call management activities, as some authors have explicitly recognised.[2] To appreciate this, an idea of the domain of management is needed.

Chapter objectives

By the end of this chapter the reader will understand:

1   The nature of management and its multiple applications for defence in the UK;
2   Defence as a broad discourse of management thinking;
3   The ideas associated with New Public Management and their applicability or otherwise for the defence sector;
4   The notion of contractors on deployed operations as a management imperative;
5   Ideas relating to performance management.

Chapter structure

We start with a discussion on the nature of management within the context of defence before exploring broader narratives and discourses of the phenomenon. This opens-up a critique of the supposed New Public Management within government and its applicability for the Ministry of Defence and its key suppliers. We introduce the idea of deployed contractors to operational theatres through the lenses of so-called economy, efficiency, and specialisation, before discussing the notion of performance management within the defence enterprise.

What is management?

The conceptualisation and deconstruction of management, as opposed to its long-standing practice, was a twentieth-century phenomenon that initially owed much to the writing of Henri Fayol,[3] a French mining engineer whose work published in French during the First World War initially attracted little attention. When translated into English just after the Second World War, that changed markedly, and his categorisation of the basic roles of the 'manager' has stood the test of time. Fayol found that a manager's job was to forecast and to prepare his organisation for future survival and success. There followed the obligation to provide direction and instruction to the workforce. Implementation of that direction involved the need to put in place the right organisational structures and to allocate individuals and groups to tasks and functions. Having divided up the enterprise into sub-units, there was also a need to promote the coordination of those sub-units so that they operated as a single machine. Finally, the manager had to keep track of how everything was going and to make adjustments to deal with emerging problems.

In the defence domain, the centrality of these tasks (forecasting, directing, organising, staffing, coordinating, and monitoring) is obvious.

- Forecasting
  There are few more ambitious efforts to look into the future than the UK's Global Strategic Trends exercises.[4]
- Direction
  As for 'direction', as well as the military ranks and discipline system, with its emphasis on unity of command, the roles of defence policy and planning documents include both the assessment of challenges and the approach to how they are to be dealt with (see previous chapter).
- Organisation
  On the organisation front, staffing, the recruitment, preparation, and retention of the right people, and placing them in organisational structures, is central to defence capability. The nature of organisational choices can sometimes provoke controversy: in 1918 the British opted to create a Royal Air Force separate from the British Army and Royal Navy,[5] whereas the United States ended the Second World War with the US Army Air Force,[6] a very different option. When the UK defence establishment decided at the turn of the millennium to acquire an attack helicopter (the Apache), the conclusion was that it should be located within the Army[7] whereas India, acquiring the same system, will base its machines in the Air Force.[8]
- Coordination
  A continuous challenge has been to discern the extent to which the branches of the armed forces should be coordinated and how that should be achieved.[9] The UK only introduced an overarching Department of State, the Ministry of Defence, after 1947, having won the Second World War using coordinating committees linking the War Department (Army), the Admiralty, and

the Air Ministry, overseen by the prime minister.[10] Moreover, the actual powers of the MoD and the responsibilities of its ministers have constantly evolved, and since the Levene Report[11] of 2011 the Joint Force Command has merited, but not necessarily achieved, a significant central role in bringing together all elements of UK military capability.

• Monitoring
  Finally, monitoring defence performance is desirable but difficult, as other parts of this chapter will demonstrate. While military operations provide a significant test of how defence is and has been managed, what is effective in one operation can prove less successful or even counter-productive in a different context with a different adversary. Also, significant and testing military operations are not a constant in defence life. In addition, while the MoD in the UK is regularly accused of 'wasting money' in the media[12] and sometimes by parliament,[13] trying to pin down precisely what waste means is often difficult. The MoD's own assessment of its performance is to be found primarily in the Annual Report and Accounts of the Ministry,[14] while the House of Commons Defence Committee supported by the National Audit Office (NAO) provide regular external assessments.

Defence and the discourse of management

The systematic study of management, and the volume of published literature on the subject, increased rapidly after 1945 and the flow continues today. Most of that writing was about the commercial world, with the readers of publications anxious to learn from the successful and negative experiences of others. It developed its own vocabulary and discourse, some of which unfortunately appeared to be designed to impress the gullible rather than inform and clarify the curious. Some of the authors' favourite phrases and tools of twenty-first-century defence include 'Rainbow Team', Match fit, Master Data Assumptions List (MDAL), 'Near-cash', and 'heat maps'. Other gems from the MoD's website include 'paradigm shift', 'inverted success', 'honourable hypocrisy', and 'flexible growth ladder'.[15]

However, ignoring the neologisms and frequency of 'management-speak',[16] the formal study of management addressed some significant concerns pertinent to both the private and public sectors. These concerns included how people are best selected, motivated and developed; how competing desirable goods and services should be prioritised; how projects should be run; what tasks should be done inside an organisation and what should be contracted to others; how is a successful and desired change to be brought about; how are costs to be calculated; and how is performance to be assessed.

The linking of wider thought about management explicitly with the MoD was a significant contribution of Michael Heseltine, himself a successful businessman as well as a former soldier, as Secretary of State for Defence between 1982 and 1986. Heseltine pioneered the allocation of 'Top Level Budgets'[17] to specific individuals within defence, with the people concerned being held responsible for the generation of 'outputs' with the funds they received. He also increased the stress on the role of information generation to keep track of what was going on and his MINIS reporting arrangements were considered radical in their time.[18] Indeed it would be fair to say that Heseltine placed the concept of 'value for money' at the centre of defence, leading to a specific public document in which the MoD sought to explain the dimensions of value.[19]

In the field of defence procurement, considered separately elsewhere in this book, Heseltine, together with Prime Minister Margaret Thatcher, was responsible for increasing the incidence of competitive procurement approaches in order to achieve better 'value for money'.[20] Peter Levene, now Lord Levene but then a private sector businessman, was brought in from outside government to head the MoD's Procurement body at a salary above civil servant grades to implement the new stress on competitive tendering.

Defence management, 'jointery' and 'capability-based' arrangements

With the Cold War over and Kuwait successfully liberated in 1991, the UK did not immediately undertake a major defence review to specify the UK's future challenges, its role in the world, and the missions expected to undertake. As noted earlier, the suspicion must be that the previous review, in 1981,[21] which had recommended significant cuts to the Navy, had caused so much controversy and division, not least within the ruling Conservative Party, that there was no appetite for a repeat exercise. What the MoD did do was accept that it would receive less funding and focus on how it could best adapt with minimum damage to front-line capability. The results were the 1990 Options for Change[22] document and the 1994 Defence Costs Study,[23] sub-titled Front-Line First, in which the MoD sought to introduce significant efficiency savings, in some cases by taking activities that had previously been conducted separately by the three services and bringing them together to achieve economies of scale. Cases that were not entirely without sensitivity included joint advanced staff training (a single Advanced Command & Staff Course), much integration of medical support, and the three service musician training schools being merged into a Defence School of Music.

The organisational trend towards more activities being managed on a tri-service, 'purple'[24] basis was continued by the Strategic Defence Review (SDR) of 1998,[25] which in policy terms confirmed the UK's ambition to be a major military player on the international stage (see the chapter on Policy) and in capability terms meant that all the elements for force projection for a range of missions had to be generated.

Although the evidence is not formally available in public, there is little doubt that the Labour government that took over in 1997 inherited a 'bow-wave' of increasing future spending commitments in the existing defence plans and costings. There was thus a need to increase both the relevance and effectiveness of UK forces by cutting their focus on defence against Moscow and building the capacity for force projection. This was coupled with a drive to increase the efficiency of British defence by changing wasteful structures and processes.[26]

Supporting the need for increased relevance, effectiveness, and efficiency were a number of organisational and wider management changes contained in the SDR 1998. Perhaps most striking, and in the face of reported military opposition, the three service logistics functions were significantly cut and a single Defence Logistics Organisation (DLO) was created to deal with the purchasing, storage, and transport associated with the support of both equipment and people. Inter alia, this much facilitated the rationalisation of storage facilities and transport arrangements.

A further step was the creation of a Permanent Joint Headquarters[27] to command current military operations from the UK and plan for potential operations. This system replaced the earlier practice of creating a specific headquarters for each operation located within the lead service for that activity. Thus, one institutionalised centre with appropriate information, communication, and command facilities could be established.

A third management feature of the 1998 SDR was its inclusion of the Smart Procurement Initiative,[28] which is analysed in more detail elsewhere in this volume. However, it is appropriate here to note the attempted transformation of the requirements generation process.

Before 1998 the specification of requirements for future systems was undertaken in the Main Building of the Ministry of Defence in an ostensibly 'purple' structure. However, the head of that structure, Deputy Chief of the Defence Staff (Systems), had underneath him four two-star Directors of Operational Requirements (for Land, Sea, Air, and Information Systems). Three of these were clearly single-service focused and took their direction from their single-service organisations located elsewhere in Main Building.

After 1998 an alternative structure was put in place headed by a three-star Equipment Capability Customer (ECC)[29] whose focus was to be on the required capabilities needed for the future rather than the specific systems or platforms needed to provide that capability. Requirements were to be written in terms of required capability rather than technical specifications. The sub-organisations within the ECC organisation were initially designed to deal with systems associated with more than one service, emphasising the novel and joint nature of the structure.

The version of British Defence Doctrine published in 2001[30] included a capabilities framework which asserted that there were seven basic military capabilities: they involved the capacities to *prepare* and train; to *deploy* to the area of an operation; to *inform* with pertinent information; to *command* through decisions and their communication; to *operate* with a possibly wide range of activities, including patrolling, suppressing, destroying, etc; to *sustain* activities over a period; and to *protect* forces against adversary efforts. Clearly, major platforms could focus largely on specific capabilities or be able to generate contributions in a range of them, but the framework brought a degree of structure to what otherwise might be seen as a rather amorphous concept.

Somewhere between driven by and endorsed by the McKinsey studies[31] underpinning Smart Procurement as a whole, the ideas of capability-based acquisition reflected at least three considerations:

Replacement thinking, which might lead an army to reason that it needed a new tank because its existing tank was getting old, was becoming ever more expensive. As had been noted even in the 1980s by the American industrialist, Norman Augustine, a replacement platform tended to cost two or three times more than its predecessor.

Replacement thinking could also lead to missed opportunities in an age where many areas of technology were moving quickly, not least in the commercial world. This was particularly the case in the area of surveillance capabilities (where radars, optical systems, infra-red technology, night vision systems, satellites, aircraft, unmanned air systems, land and sea-based systems were all relevant). But it was also clearly a factor to be taken into account in the field of precision-strike, for instance, where land, air, and sea-based missiles using a flight (cruise) or ballistic trajectory were alternatives to long-range artillery and gravity-bombs.

Finally, after the end of the Cold War, Western countries could not be confident about the adversaries they were likely to face, indeed the only predictable element was that Western forces were likely to be taken by surprise about where they would be asked to operate next. In these circumstances, there was no specific enemy against which threat-based requirements could be written. A broader approach that focused on generic capabilities provided a useful handrail in the absence of a benchmark adversary.

In practice, this capability-based system had a hard time for essentially bureaucratic political reasons: the single services continued to want a major voice in future requirements and their prioritisation. In addition, to provide a hypothetical illustration, the Royal Artillery which operated guns could not be expected to be enthusiastic about the Air Force taking on some of their responsibilities by the procurement of air-ground missiles. An officer appointed to a joint post in the ECC still depended on his own service and even service branch for his/her next post and any promotion.

However, the Smart Procurement Initiative, when broadened to Smart Acquisition[32] from around 2003, resulted in capability development being placed on a much more systematic basis. Particularly in the light of issues arising with the introduction into service of two systems novel to the Army (the Apache attack helicopter and the Bowman tactical communications system[33]), the TEPIDOIL framework was developed under which capability was underlined as not being limited to the delivery of 'equipment': it needed also the systematic and coordinated generation of the right training, people, infrastructure, doctrine, organisation, information and logistics. Management responsibility for monitoring the timely development of these elements was eventually entrusted to 'Senior Responsible Owners'[34] (SRO) in the MoD or single services, and the National Audit Office also included an assessment of the risk associated with these factors in its annual reports on major equipment projects.[35] However, understanding the TEPIDOIL framework was no assurance that its elements would be delivered.

Although the government had agreed to the development and building of two aircraft carriers in 1998, and although it took almost 20 years for the first to be delivered for sea trials (in 2017),[36] *The Times* reported in February 2017 that 'there is concern that the manpower shortage in the navy is so great that it will not be fully able to staff the new carriers and the rest of the fleet'.[37]

Managing money: Resource Accounting and Budgeting (RAB)

A massive change to defence management came about in the latter part of the 1990s as a result of the then Conservative administration's decision to adopt a version of commercial accounting across government. This was the Resource Accounting and Budgeting (RAB)[38] initiative of which the main anticipated gain was to generate a much better sense of the cost of activities and capabilities, which in turn should generate better choices and priorities.

Prior to the introduction of RAB, the MoD, like the rest of government, used a system of cash accounting. This simply kept track of what had been spent and on what, but it had great limitations in dealing with capital expenditures. Capital spending comprises one-time expenditures on goods with the expectation of the benefits flowing from those goods being gained over a significant number of years ('significant' normally being envisaged as a minimum of five years). When the government pays for the construction of a hangar to store and repair aircraft, it might envisage that hangar being useful for 30 years. The Trident submarines, which were delivered and paid for from 1984, are to be in service, i.e. useful, until 2031.

Under cash accounting, capital spending appears as a single lump sum of which there is no trace in succeeding years because the money has been expended. Thus, if the question is posed, how much does it cost to operate the nuclear deterrent fleet, the honest answer under cash accounting would have to be that it would depend on which year or years were examined. In the early years, when the submarines were being built and paid for, the cost would seem hugely high. Then, afterwards, no capital spending would be needed, it would appear remarkably low with much of the relevant costs just arising from the salaries of crews. Salaries and other spending on items such as fuel are 'recurrent' expenditures because the benefits associated with them are used up quickly and more money needs to be spent. Strictly, in the case of salaries, the employer has already received the benefit of the payment (the work of the employee) before the payment is made.

Private companies, preparing accounts to indicate the state of their financial health, treat capital spending separately, and record as costs the 'resources' of the original capital spending used up in a single year. In other words, they measure the 'depreciation' of their assets. If a factory buys looms to weave cloth over a decade, they would charge one tenth of the resources associated with the looms' purchase to the factory's profit and loss statement for the year.

In moving from cash accounting to RAB, the MoD found it significantly easier to calculate the costs, defined in terms of resources used, of specific elements such as the Apache fleet or a Typhoon squadron. However, implementation of RAB in practice was a matter of significant complexity and challenge for the MoD.

This was because, uniquely among government departments, the MoD owned a huge number of capital assets in the form of equipment, infrastructure, and stocks, all of which had to be valued according to (international) accounting rules so that an MoD 'balance sheet' of assets and liabilities could be prepared. Ministries such as Health and Education, while overseeing many capital assets, did not own them because ownership was vested in the delegated regional and legal authorities. Other ministries thus had a much easier time. In defence, there were numerous judgements to be made, for instance about which spending on research and development should be treated as recurrent spending and which could be classed as capital expenditure. The MoD also owned large stocks of spares and ammunition for specific systems which needed to be classified and valued according to an accepted accounting principle.

The introduction of RAB was spread over several years and was far from straightforward. Apart from any other consideration, the MoD had to recruit and train more than 1,000 qualified accountants to implement the system.

One result of RAB, and the specific accounting concepts and terms it required, is that the MoD's Annual Report and Accounts are largely unintelligible to the layman, including many members of parliament who are expected to oversee the MoD. There might be reasonable suspicion that many military personnel and civil servants only have a limited understanding of how RAB works. Specifically, it often requires very careful investigation to discern whether a number refers to cash actually expended in a period (which parliament still allocates) or whether it refers to resources used up in a period.

New Public Management

From the time of Mrs Thatcher as prime minister, the British government had been drawn to many of the ideas of the literature of New Public Management (NPM), and the MoD was not exempt from this.[39] NPM expressed strong faith in competitive market structures as a means of getting suppliers to deliver value to customers, with the government seeking to apply this logic to schools and hospitals. There was an emphasis on formal performance management against specified targets and, crucially for defence, an expectation that private sector bodies were almost inevitably more efficient at the delivery of goods and services than public sector bodies, especially if competitive pressures could be used.[40] This led to the outsourcing of many activities that had previously been undertaken by public servants.[41]

Most obviously in British defence it led to the privatisation of nationalised businesses, including British Aerospace,[42] Rolls Royce,[43] British Shipbuilding,[44] the Royal Ordnance Factories,[45] and the remaining naval dockyards.[46] In 2001 the Defence Evaluation and Research Agency (DERA) was split and much of its content passed to a private company that called itself QinetiQ.[47] In 2015 Defence

Support Group, a maintenance, repair, and refurbishment organisation dealing with armoured vehicles and aircraft, was sold to Babcock.[48]

Outsourcing affected almost every facet of defence, including facilities management,[49] training (contractors were used to maintain training areas and design exercises),[50] education (involving Cranfield University and Kings College as far as higher education was concerned),[51] recruitment (where Capita was given responsibilities),[52] and equipment support.[53] In the latter area companies were increasingly contracted to undertake maintenance, repair, and other activities against a requirement to provide availability rates for equipment. Previously their responsibilities had been limited to the assured and timely supply of spare parts. Contracting for availability,[54] it should be noted, incentivised firms to modify equipment to make it more reliable and less in need of support work. Companies that solely sold spares made more money the more spares were needed!

The Private Finance Initiative (PFI) constituted a particular form of outsourcing that was widely adopted in British defence. The essence of a PFI is that the government, instead of buying a capital asset and then generating services and value from it, contracts with a company for a capital-based service or services. The company arranges and pays for the generation of the capital asset (which is depreciated across the life of the PFI), provides the labour and other materials needed for the delivery of the service(s), and charges the government (monthly) for what is delivered.[55]

Two contrasting PFIs were for the Advanced Command and Staff College with associated accommodation at Shrivenham and the Voyager fleet of tanker/transport aircraft for the RAF.

In the case of the staff college, a consortium of Laing-Serco (now just Serco as Laing sold its stake) committed to build a staff college and associated student and staff housing, and then to provide housekeeping, catering, educational, and other services for 30 years so that MoD staff and students could use the facilities in exchange for a monthly payment.

With Voyager ('special purpose' vehicle called AirTanker),[56] a consortium of companies led by Airbus contracted to provide and support a fleet of 12 aircraft which could be used in tanker and other roles by the RAF crews that actually fly the aircraft. The arrangement was and is that the RAF agreed to use eight of the aircraft on a constant basis while AirTanker was free to seek other customers in the civil or military world for the remainder of the fleet. However, in cases of military need, Airbus must be able to call back the whole fleet and make it available to the RAF.

The advantages for the MoD here were that it did not need to fund the capital cost of the aircraft. This cost was recovered by the firm through the monthly payments. It also did not have to calculate the annual costs of operating the fleet: the bidding companies did that. Finally, it could hope that, through the private work that Air Tanker could generate, it might obtain a lower cost for the service than it would by buying the aircraft itself. The jury is out on this last point.

From the mid-1990s the MoD agreed a very large number of Private Finance Initiatives, including the Military Flying Training System[57] arrangement which entrusted to the private sector much of the training of both fixed wing and heli-copter pilots. By 2016 the MoD had more than 40 PFIs where it was the exclusive user of the service involved, plus a few more, including Voyager, where there was some element of third-party use and thus revenue for the contractor. While these had enabled the MoD to avoid major lumps of capital expenditure at a time of scarce resources, they also meant it was committed to significant multi-year spending.

PFIs had a notable link to RAB and international accounting rules. Originally it had been hoped that capital expenditure on PFIs could be kept off the MoD and government balance sheet because, after all, the money had been spent by the private sector. Thus, the capital sums would not count as an element in the national debt. However, it transpired that only when there was significant third-party use could this apply. When the MoD was the only user of a capital-based service, the capital costs had to be recorded on the MoD's books. This ruling significantly reduced the attractiveness of PFIs, but the MoD's enthusiasm for outsourcing to the private sector through service contracts remained significant.

Contractors on deployed operations: the beginning of the 'Whole Force'

A notable and somewhat controversial area of outsourcing concerned the use of contractors on deployed operations.

UK forces conducted the first Gulf War in 1991 with minimal use of contractors.[58] Operation Herrick in Afghanistan, 25 years later, was run with about eight contractors in theatre for every ten British soldiers.[59] The United States actually ended the campaign in 2014 with more deployed contractors than soldiers.[60]

This came about largely on an unplanned basis and reflected the reality that, while the MoD had 'managed' the generation of the capacity to deploy trained and equipped forces on a specified scale, it had not prepared carefully for the unexpectedly protracted and difficult operations that occurred to a degree in the Balkans and emphatically in Iraq and Afghanistan. Thus, to an increasing degree, the MoD used contractors for an ever-wider range of activities, including infra-structure construction and support, people support (covering catering, personal hygiene facilities, off-duty entertainment, and communications), transport of supplies, and equipment support.[61] Although UK contractors were limited to sup-porting operations rather than undertaking aspects of operations (with the excep-tion of interpreting and translation services), this was not true of the US[62] and even the Netherlands.[63] The latter's Hermes surveillance UAVs were operated by Thales staff.[64] Moreover, even the UK had to use armed contractors to protect UK diplomats and other government agents in hazardous theatres, if only because the British military lacked the numbers of personnel to undertake these duties.

By the end of 2014 and the conclusion of Operation Herrick in Afghanistan, the MoD had formal guidance on the use of CONDO (JSP 679) and had coined the term Whole Force[65] to capture the character of deployable force for the foreseeable future. The Whole Force comprises full-time military, civil servants, reservists including sponsored reserves, and personnel from the private sector. The MoD Joint Service Publication 1-05, Personnel Support for Joint Operations, included the following statement in 2016:[66]

> 1.7. The term 'Whole Force concept' has been in existence for several years and its fundamental principles have influenced the evolution of the force mix to the extent that it is now the time to move from a Whole Force concept to a Whole Force approach. A Whole Force approach will ensure that Defence's goals are achieved by high-performing, fully-integrated, well-led teams of capable and skilled people drawn from our Regular and Reserve UK Armed Forces, the civil service, MOD civilians and contractors. Constraints still apply to managing the Whole Force, which limits the value that can be delivered.
> These constraints include:
>
> - Political – for example, the limitations on the numbers of civil and military personnel;
> - Procedural – for example, the rules and regulations which apply to the employment of different categories of people;
> - Cultural – for example, the nature of the relationship between MoD and contractors.

Thus, in moving from a concept to an 'approach', the UK accepted that its conceptual model for the generation of military capability had to include reference to reservists and purely civilian staff from the private sector. This had clear implications for performance management and the definition of 'outputs' to be generated from the defence budget.

For some years the Permanent Joint Headquarters (PJHQ) had had a contract with a company lead on the provision of services that the MoD might need in the event of military operations and, in October 2017, the latest iteration of this contract was awarded to Kellogg, Brown and Root (KBR).[67] The wide range of services that the MoD expected the company to be able to arrange was broad and covered in the Statement of Requirement for the Operational Support Capability Contract. The committed reader can refer to the original documentation: suffice it here to note a requirements sample, including that the contractor was to be able, inter alia, to:

- Provide 'a storm debris and snow and ice clearance service for all locations';
- Provide new roads required by the deployed military staffs, for use by up to Class 120 Wheeled (120W) and 70 Tracked (70T) vehicles, and vehicles with loads. Roads may be provided using either expedient or conventional methods;
- Provide laundry, ablutions, and tailoring services as required.[68]

Performance management

A key management responsibility is monitoring how well an organisation is performing. In the commercial world, this can be challenging but, in defence, it is arguably more difficult. For five years after 1998 the UK was not conducting major military operations, the new financial system required the MoD to publish an Annual Report and Accounts in which performance measurement and management gained prominence. The focus of concern was how well the MoD was preparing capability rather than using it.

The SDR of 1998 had specified the missions and tasks that the military had to carry out and the units of forces that were to be ready for operations at different elements of readiness. Force Elements at Readiness (FE@R)[69] were the primary 'outputs' generated by much of the defence budget and the MoD's performance was defined significantly in terms of how well it had done in the generation of these outputs. Readiness was defined largely in terms of the availability of trained people and needed equipment, plus some concern with logistics.

However, the MoD was also influenced by debates about performance measurement and management in the commercial domain, where companies had realised that it was often not enough simply to measure profit at the end of the year. Companies needed to feel that they had also done what was necessary to prepare for the future, with the change and challenges that it might bring. In the business world a highly influential article (1992)[70] and then book was Kaplan and Norton's *The Balanced Scorecard* (1996),[71] which argued that there were four broad dimensions of performance. These required organisations to examine how they looked to their customers; how they looked to their shareholders and the financial world; how efficient and effective were their internal processes; and whether they had done enough learning, and developing their people and other assets, to remain competitive in the future.

The idea of the Balanced Scorecard always recognised that firms could select their own structure of performance metrics, and thus it was quite feasible that the MoD could adapt the idea that, under four broad headings, its performance could be measured using a number of indicators. Thus, there was an effort to permeate the tool through the MoD and, at the highest level, the MoD's Annual Report and Accounts used a Balanced Scorecard approach. This was meant to measure the condition of key variables and also to direct attention to where future managerial attention needed to be directed. It was also linked to risk management to signal issues of particular concern.

The focus on and domination of operations in Iraq and then Afghanistan moved attention away from the multi-dimensional approach to performance assessment, and the end of Operation Herrick by the close of 2014 did not result in the Balanced Scorecard being brought back. The MoD's Annual Reports became largely public relations documents highlighting specific items of good news, and the whole question of the actual production of defence outputs, i.e. force elements available for operations today and in the near future, largely disappeared from the public agenda. Instead, reporting was devoted to progress towards targets of the future, how well the MoD was moving towards putting in place a force for 2020 (after the SDSR of 2010)[72] and then a force for 2025 (after the SDSR of 2015).[73]

2010 and 2011: the Levene reforms

The security and defence review of 2010 was undertaken in a hurry, with a coalition government elected in April feeling the pressure to produce an affordable defence policy by the time of the announcement of the four-year Comprehensive Spending Review in the autumn. Detailed consideration of how management of the MoD was to be changed was entrusted to Lord Levene and a smaller group who reported in November 2011.[74]

The Levene Report resulted in more than 40 change workstreams but at their essence was the aim to delegate financial responsibility to the four commands (Land, Air, Sea, and Joint Forces). In terms of the costs of equipment support, the change was unambiguous: before 2011 the single services had specified their use needs for equipment, and Defence Equipment and Support (DE&S) then calculated the support costs involved, and worked out what could be afforded. The head of DE&S was the 'Top Level Budget' (TLB) holder for equipment and support purchases. After 2011, the single services and Joint Forces Command (JFC) TLBs included the money for equipment support, so they had to manage funding within the constraints of the money available. DE&S became the chief negotiator with the private sector and the provider of information about costs to the commands. For the Army in particular, with its multitude of equipment types with interlocked cost structures, the financial management of support was a major challenge.

On the purchase of new equipment, service responsibility was more qualified. The service budgets included the capital funding for new requirements but the capital sums allocated to each command came from the Finance and Military Capability (FinMilCap) section in the MoD centre, which also led the approval processes associated with suggestions for new equipment and also managed 'strategic programmes' such as the deterrent, complex weapons, and for an extended period the carrier-strike programme. In short, on new equipment, the delegation of powers to the single service and Joint Force Command was significantly qualified.

Management and the viability of policy ambition

Echoing the SDR of 1998, the sub-reviews of 2003 and 2004 and the major reviews of 2010 and 2015 have specified that the UK would be centrally involved in supporting security in the wider world, significantly by being a major military power. According to these, this involved the re-orientation and even restructuring of UK defence forces, for instance increasing the need for global communications and the transport by air and sea of UK forces, their equipment and their supplies, notably specifying 'defence outputs', output being a managerial term, in the form of force elements at specified rates of readiness.

The government's calculations showed that, if the MoD's organisation, manpower, processes, and so on stayed as they were prior to 1998, there would not be enough money to generate the outputs specified. Thus, the government committed to increasing the efficiency of the MoD, i.e. improving its management, in order to free up resources that could be used to deliver the policy. This reliance on management improvements to enable the delivery of an ambitious policy has thus been a persistent feature of reviews in the past 20 years.

Thus, the status of management as a field of activity moved from just being the means of organisation, coordination, and the means through which policy was implemented to being the essential enabler of making a policy affordable within the limited resources available. In the SDSR of 2015, the MoD committed itself to savings of £9.1 billion over a five-year period in order to render the equipment plan and other programmes affordable.

Chilcot

The massive Chilcot Report into the initiation and delivery of the 2003 campaign in Iraq published in 2016[75] contained many criticisms of the MoD and represented the most recent input into MoD management reform. Inter alia, it raised many doubts on the efficacy of the scheme for the generation and delivery of Urgent Operational Requirements (UoR)[76] to fill capability gaps, a system that at the time had been widely thought to have worked well.

In 2016 a small team had been established in the MoD, looking to establish a programme of change to respond to its findings. It developed a programme focused on three areas: improving the governmental defence sector's ability to identify, curate and access lessons from projects and operations; strengthening the professional knowledge base of defence people, especially those concerned with security policy and operations; and, arguably most ambitious, building a challenge culture in the MoD in which people at all levels would feel free to offer new thinking.

Conclusion

It can be argued that, for more than the last 20 years, supposed management improvements have been central to UK efforts to ensure that tangible defence capabilities of a specified character were put in place as part of the implementation of defence policy. Such improvements were also seen has necessary if the policy aspirations of the UK government to generate defence forces capable of major operations on the global stage were to be affordable. Moreover, efforts to improve the management of the defence sector in terms of personnel, information, finance, and equipment have been more or less continuously giving management reform, along with its nephew acquisition reform, the character of a soap opera rather than a novel with a specified end.

However, critics of progress in this field need to take into account at least six related chronic conditions that affect defence and limit what is possible.

The first is the limited possibilities for and unintended consequences of delegation in defence. Very large organisations often seek to delegate responsibilities and powers down to sub-sections, knowing that people are more motivated and work better when they have some discretion about how they execute their duties. Defence recognises this, most obviously by breaking itself into three services and today four commands, through its top-level budget system and indeed through the frequent application of mission command thinking. However, the expectation of the armed forces and their support groups is that they constitute one single well-coordinated and integrated, near organic machine. Delegation and integration are rarely easy to reconcile, as the experience of running integrated project teams under the Smart Procurement Initiative demonstrated. The links among related projects led to friction and, by the time the coalition government came to power in 2010, the recognition was in place that projects needed to be coordinated at the programme and even portfolio levels.

The second element is that defence is required to operate at very different rates. In peacetime the military are largely concerned with training. This is largely a planned activity whose costs can be calculated and managed so as to be minimised, for instance so that inventory management delivers 'lean' and 'just-in-time' supplies of parts. Training on an increasing number of systems, not just aircraft, is dominated by the use of simulators rather than the actual system of concern in order to save money. In contrast, when military operations occur, which since the end of the Cold War have involved a degree of surprise, usage rates of equipment rise and the 'just-in-case' availability of parts and munitions becomes much more prominent in the minds of users. In these circumstances, when the defence sector is looking for efficiency savings, should the focus be on peacetime efficiency or 'wartime' efficiency when capability shortfalls can result in loss of life and political setbacks. Engineers tell us that physical machines maximise their efficiency within a specific performance range. The commercial jet engine has maximum fuel efficiency when cruising at altitude. The automobile engine is optimised for fuel consumption and wear and tear when driving at a consistent 55 miles per hour. A much more complex human-machine system such as the governmental defence sector cannot be expected to be ideally efficient in both training and operational contexts.

A third chronic condition with which defence must live is that the world can change quicker than defence can. The collapse of the Soviet Union, the invasion of Kuwait in 1990, the attacks of 11 September 2001, and perhaps the election of President Trump were sudden developments of fundamental importance for defence. But adjusting defence to fit a new world takes years, and not just because new defence equipment takes so long to develop and produce. New circumstances can demand different sorts of people, novel training programmes, new doctrine for new missions, and perhaps organisational adjustments. Defence is often likened to a super tanker that takes time to change course and the extent to which defence can and wants to use Risk Management literature to prepare for change is a matter for debate.

Fourth, many organisations recognise the value of having their staff driven by some higher purpose than simply the earning of financial reward in terms of salary and/or businesses. People are recognised as performing better when more than 'transactional' considerations are in play. Those in uniform, whether full-time or reservists, are expected to put their lives at risk to a much greater extent than any other workers in Western societies. They are expected to do this in part in exchange for money, but significantly also because of a sense of duty to the higher cause of 'Queen and country'. Civil servants are expected to show a similar emotional commitment to the national interest as defined by the government, although they are not expected to put their lives on the line. But what of the relationship to defence of the many individuals and organisations in the private sector that support defence. Since more than half the defence budget is spent with the non-governmental sector, much of which may not be British, it might reasonably be expected to generate more than half the elements of capability associated with defence. Thus, a persistent management challenge in defence is to calculate the right balance of transactional and emotional drivers in the defence machine, not just with regard to government employees but also the private sector that plays an increasing defence role.

Fifth, the military sector is pressed towards perhaps excessive optimism and a can-do culture, and there are two different reasons for this:

- Most obviously, the military do not look for those driven by doubt and a lack of confidence for their ranks: they look for optimistic problem solvers. The last thing that a unit needs on a demanding operation is a majority of officers who constantly see only difficulties.
- Second, the military can be drawn to opportunity to show their relevance and utility because the British public do not directly experience the benefits of the defence sector in normal times. This is in contrast, for example, to the services provided by the education and health sectors. The US academic Joseph Nye captured this phenomenon with his observation that 'Security is like oxygen–you tend not to notice it until you begin to lose it, but once that occurs there is nothing else that you will think about'.[77] Even when the military can claim to be delivering a deterrence function, they are largely invisible to the public and, especially in the post-Cold war era, many military units are in simple terms paid to do nothing, but to be ready to do difficult things at short notice. Arguably this all puts military leaders under constant pressure to embrace with public confidence any prospective task that is put before them as a demonstration of their relevance. On occasions the UK benefits significantly, as when UK troops provided short-notice security for the Olympics,[78] and at other times things go significantly wrong. The Army's enthusiasm to take responsibility for Helmand in 2007 is a case in point.

Sixth, defining success has become very difficult in the nature of many of the operations that today's military are called on to undertake. Unambiguous victory over a defined and structured opponent is rarely feasible and then the translation of positive military results into long-term political consequences often proves demanding. Interventions such as those in Yugoslavia, Iraq, Afghanistan, and Libya were about first using Western forces to end or at least restrict internally-generated violence, but then there was an expectation that UK and allied forces would be able to promote the effective creation of stable, well-governed, and peaceful states and societies. The latter represented social engineering tasks of a most demanding nature, tasks which in some cases involved resolving problems that were centuries old.

These six challenges are not issues that can be resolved in some finite sense. They have to be accommodated, lived with, and minimised in their impact.

Moreover, in defence there is an elusive right balance of the effective exploitation of wider management thought and literature, and the excessive use of what military officers can easily view as meaningless management speak. Some management ideas and major works have demonstrated their utility in the defence sector, but the MoD has not entirely lost its appetite for the glib but empty phrases. Two decades of study have led these authors to the conviction that many sub-areas of management study, including Human Resource Management, Performance Management, Change Management, Portfolio, Programme and Project Management, Logistics and Supply Chain Management, Strategy and Risk Management, can throw much light on the challenges found in defence, but they must be used in conjunction with awareness of why and how defence is often different from the commercial world and even many other areas of government.

Notes

1 Evident in the officer recruitment webpages of each of the single services. See: www.army.mod.uk/join/Join-as-an-Officer.aspx; www.raf.mod.uk/rafregiment/careers/ officers.cfm; www.royalnavy.mod.uk/careers/joining/get-ready-to-join/royal-marines-commando-officer/whats-the-joining-process.
2 See John Louth, *A Low Dishonest Decade: Smart Acquisition and Defence Procurement into the New Millennium* (Cardiff: UWIC, 2010).
3 Dunod et Pinat, *General and Industrial Management* (1917).
4 The latest volume, Ministry o Defence (2014), *Strategic Trends Programme: Global Strategic Trends – Out to 2045*, Fifth Edition, to be found at www.gov.uk/government/ publications/global-strategic-trends-out-to-2045.
5 John Sweetman, 'Crucial Months for Survival: The Royal Air Force, 1918–19', *Journal of Contemporary History,* 1984, 19:3, pp. 529–47.
6 Wesley F Craven., James L Cate, Arthur B Ferguson, John E Fagg, Joseph W Angell, Alfred Goldberg, Robert H George, Albert F Simpson, Robert T Finney, Harris Warren, David G Rempel and Martin R Goldman (1983), *The Army Air Forces in World War II*, Office of Air Force History, Washington DC.
7 Ministry of Defence, *Building an Air Manoeuvre Capability: The Introduction of the Apache helicopter*, HC1246, 31 October 2002.
8 Franz-Stefan Gady, 'India Clears Purchase of 6 US-Made Attack Helicopters', *The Diplomat*, 18 August 2017, to be found at https://thediplomat.com/2017/08/india-clear-purchase-of-6-us-made-attack-helicopters/.

9 See, for examples, Andrew Dorman, Mike Smith and Matthew Uttley (eds), *The Changing Face of Military Power: Joint Warfare in an Expeditionary Era* (Basingstoke: Palgrave Macmillan, 2002); Trevor Taylor, 'Jointery: Military Integration', in Teri McConville and Richard Holmes (eds), *Defence Management in Uncertain Times* (Psychology Press, 2003), pp. 70–89.

10 Ministry of Defence, *Appraisal Report: Ministry of Defence and the Armed Forces 1963–2014*, 31 July 2014.

11 Ministry of Defence, *Defence Reform: An independent report into the structure and management of the Ministry of Defence* (London: The Stationary Office, 2011).

12 See, for example, BBC News, 'MoD "wastes money storing unnecessary supplies"', 28 June 2012; Jeremy Warner, 'The MoD is still losing the war on waste', *Telegraph*, 9 June 2016.

13 See, for example, Richard Norton-Taylor, 'MPs attack MoD over £8bn weapons project waste', *Guardian*, 22 February 2011; House of Commons Defence Committee (2015), *Decision-Making in Defence Policy*, Eleventh Report of the Session 2014–15.

14 The latest version, Ministry of Defence, *Ministry of Defence Annual Report and Accounts 2016 to 2017*, HC 21 2017–18, to be found at www.gov.uk/government/publications/ministry-of-defence-annual-report-and-accounts-2016-to-2017.

15 These terms can be found on various pages and documents published on www.gov.uk/government/organisations/ministry-of-defence.

16 Andrew Brighton, Introduction – 'Management Speak: a master discourse', *Critical Quarterly*, 2002, 44:3, pp. 1–3; David Greatbatch and Timothy Clark, *Management Speak: Why we listen to what Management Gurus tell us* (Oxon: Psychology Press, 2005); Sally Babidge, Shelley Greer, Rosita Henry and Christine Pam, 'Management Speak: Indigenous Knowledge and Bureaucratic Engagement', *Social Analysis*, 2007, 51:3, pp. 148–64.

17 Ministry of Defence, *Appraisal Report: Ministry of Defence and Armed Forces 1963–1914*, 1 July 2014, p. 13.

18 Ewan Broadbent, 'Military and Government: From Macmillan to Heseltine', *RUSI Defence Studies Series*, 1988, pp. 59–83.

19 Ministry of Defence, *Value for Money in Defence Equipment Procurement*, Defence Open Government Document 83/01, 1983.

20 Andrew Doorman, *Defence under Thatcher* (Basingstoke: Palgrave Macmillan, 2002).

21 Ministry of Defence, *The UK Defence Programme: The way forward*, CM8288, 1981.

22 International Security Information Service, *Options for change: the UK defence review, 1990–91*, 1991, p. 21.

23 Tom Dodd, *Frontline First: The Defence Costs Study*, Research Paper 94/101, 1994.

24 Sherard Cowper-Coles, 'From Defence to Security: British Policy in Transition', *Survival*, 36:1, 2008, pp. 142–61.

25 Ministry of Defence, *Strategic Defence Review: Modern Forces for the Modern World* (London: The Stationary Office, 1998).

26 Michael Codner and Michael Clarke, *A Question of Security: The British Defence Review in an Age of Austerity* (London: IB Tauris, 2011).

27 Martin Edmonds, 'Defence Management and the impact of Jointery', *Defense Analysis*, 14:1, 1998, pp. 9–27.

28 Colin McInnes, 'Labour's Strategic Defence Review', *International Affairs*, 1998, 74:4, pp. 823–45.

29 Ministry of Defence (1998), op. cit.

30 Ministry of Defence, *British Defence Doctrine*, JWP 0-01, 2001.

31 Ministry of Defence (1998), op. cit., p. 12.

32 Claire Taylor, *UK Defence Procurement Policy*, Research Paper 03/78, 2003 pp. 16–46.

33 National Audit Office, *Ministry of Defence: Delivering digital tactical communications through the Bowman CIP programme*, HC 1050 Session 2005–2006, 2006.

34 See, www.finance-ni.gov.uk/articles/roles-and-responsibilities-senior-responsible-owner for a description of the roles and responsibilities of SRO's.

35 National Audit Office, *Ministry of Defence: Major Projects Report 2003*, HC 195 Session 2003–2004, 2004.

36 S Osborne, (2017), 'HMS Queen Elizabeth aircraft carrier sets sail for sea trials', *Guardian*, 26 June 2017.

37 Sean O'Neill, 'Desperate navy seeks senior servicemen', *The Times*, 13 February 2017, www.thetimes.co.uk/article/desperate-navy-seeks-senior-servicemen-5tznx96zf, accessed 11 July 2017. The piece reported that retired sailors would be able to sign up as full-time reservists on five-year contracts with a retirement age of 60.

38 Angela Gillibrand & Brian Hilton, 'Resource Accounting and Budgeting: Principles, Concepts and Practice—The MoD Case', *Public Money and Management*, 1998, 18:2, pp. 21–8.

39 John Louth, *A Low Dishonest Decade: Smart Acquisition and Defence Procurement into the New Millennium*, (Cardiff: University of Wales Institute, 2010); John Louth and Rebecca Boden, 'Winging It? Defence Procurement as Risk Management', *Financial Accountability & Management*, 2014, 30:3, pp. 303–21; Lauren Twort, *Peace and Recovery: Witnessing Lived Experience in Sierra Leone*, PhD Thesis, Business School University of Roehampton, London, 2015; Gabriela Thompson, *For Effect or Affect? UK Defence Change: Management*, PhD Thesis, University of Roehampton, 2017.

40 JFFE De Araujo, 'Improving Public Service Delivery: The Crossroads between NPM and Traditional Bureaucracy', *Public Administration*, 2001, 79:4, pp. 915–32.

41 Paul Bishop, 'Competition and Collaboration in the Provision of Public Services: The Case of the UK Defence Sector', *Journal of Finance and Management in Public Service*, 2003, 3:1, pp. 13–24; Matthew Uttley, *Contractors on Deployed Military Operations: United Kingdom Policy and Doctrine*, Strategic Studies Institute, 2005; Christopher Kinsey, *Private Contractors and the Reconstruction of Iraq: Transforming Military Logistics*, (London: Routledge, 2009); Henrik Heidenkamp, 'Sustaining the UK's Defence Effort: Contractor Support to Operations Market Dynamics', *RUSI Whitehall Report 2–12*, 2012; Henrik Heidenkamp., John Louth and Trevor Taylor, *The Defence Industrial Triptych: Government as customer, sponsor and regulator*, *RUSI Whitehall Paper* 81, 2013; Thomas Ekström and Michael Dorn, 'Public Private Partnerships in Defence Acquisition', in Eβig and Glas (eds), *Performance Based Logistics*, Springer, 2014, pp. 303–24.

42 David Parker, *The Official History of Privatisation Vol. I: The Formative Years 1970–1987* (London: Routledge, 2009) pp. 113–19.

43 Ibid., 2009, pp. 198–201.

44 Lewis Johnman, 'The Privatisation of British Shipbuilders', *International Journal of Maritime History*, 1996, 8:2, pp. 1–31; David Parker, *The Official History of Privatisation Vol. I: The Formative Years 1970–1987* (London: Routledge, 2009), pp. 202–3.

45 National Audit Office, *Ministry of Defence: Sale of Royal Ordnance Plc*, HC 352 Session 1987–1988, 1987.

46 David Parker, *The Official History of Privatisation Vol. I: The Formative Years 1970–1987* (London: Routledge, 2009), p. 232.

47 House of Commons Committee of Public Accounts, *The Privatisation of QinetiQ*, Twenty-Fourth Report of Session 2007–08, HC 151.

48 Babcock International, 'Babcock selected as preferred bidder for the acquisition of Defence Support Group', 19 November 2014.

49 Defence Infrastructure Organisation, 'MOD unveils facilities management contract', *News story*, 19 March 2014.

50 For example, the UK Military Flying Training System provided by Ascent Flight Training under a Private Finance Initiative contract. See, National Audit Office, *Ministry of Defence: Major Projects Report 2008: Project Summary Sheets*, 2008.

51 www.da.mod.uk/about-us.
52 www.capita.com/sectors/defence/.
53 Ministry of Defence, 'Defence Secretary announces £55m contract for UK bomb disposal robots at DSEI', *News story*, 13 September 2017.
54 Christopher J Hockley, Jeremy C Smith and Laura J Lacey, *Contracting for availability and capability in the defence environment, Complex Engineering Service Systems* (Springer, 2011), pp. 237–56.
55 Think Defence, 'MoD Private Finance Initiatives', *Think Defence*, 21 January 2015.
56 www.airtanker.co.uk/raf-voyager.
57 Ministry of Defence, '£1.1 billion investment by MOD in future military flying training', *News story*, 2 February 2016.
58 Matthew Uttley, *Contractors on Deployed Military Operations: United Kingdom Policy and Doctrine*, Strategic Studies Institute, 2005, p. 1.
59 Ibid., p. 10.
60 Heidi M Peters, Mosche Schwatze and Lawrence Kapp, *Department of Defense Contractor and Troop Levels in Iraq and Afghanistan: 2007–2017*, Congressional Research Service, 28 April 2017, p. 4.
61 Uttley (2005), op. cit.
62 Heidi M Peters., Mosche Schwatze and Lawrence Kapp, *Department of Defense Contractor and Troop Levels in Iraq and Afghanistan: 2007–2017*, Congressional Research Service, 28 April 2017.
63 Directorate-General for External Policies of the Union, 'The Role of Private Security Companies (PCSs), 2011' in *CSDP Missions and Operations, European Parliament*, EXPO/B/SEDE/FWC/2009-01/LOT6/10/REV1, p. 15.
64 Craig Hoyle, 'Thales lands support contract for Watchkeeper UAV', *FlightGlobal*, 22 March 2010.
65 Ministry of Defence, 'Partnership: taking the Whole Force Approach in defence', *Defence Contracts Online.*
66 Ministry of Defence, *Personnel Support for Joint Operations*, Joint Doctrine Publication 1-05, www.gov.uk/government/uploads/system/uploads/attachment_data/file/469880/20151014-Pers_spt_JDP_1_05_Secured.pdf.
67 KBR, 'KBR Awarded Contract to Deliver Facilities Management Services in the Middle East for UK MOD', *KBR press release*, 9 October 2017.
68 'MOD Operational Support Capability Contract', 13 November 2015, www.government-online.net/mod-operational-support-capability-contract/; 'KBR selected by UK Ministry of Defence for Operational Support Capability Contract', 10 May 2016, https://kbr.com/about/newsroom/press-releases/2016/05/10/kbr-selected-by-uk-ministry-of-defence-for-operational-support-capability-contract.
69 Ministry of Defence, *Defence Joint Operating Concept*, Joint Concept Note 1 14, pp. 3–7.
70 Robert S Kaplan and David P Norton, 'The balanced scorecard: Measures that drive performance', *Harvard Business Review*, January–February 1992, pp. 71–9.
71 Robert S Kaplan and David P Norton, 'The Balanced Scorecard: Translating Strategy Into Action', 1996, *Harvard Business Press*.
72 HM Government, *Securing Britain in an Age of Uncertainty: The Strategic Defence and Security Review*, CM7948 (London: The Stationary Office, 2010).
73 HM Government, *National Security Strategy and Strategic Defence and Security Review 2015: A Secure and Prosperous United Kingdom*, CM9161 (London: The Stationery Office, 2015).
74 Ministry of Defence, *Defence Reform: An independent report into the structure and management of the Ministry of Defence* (London: The Stationary Office, 2011).
75 Sir J Chilcott, *'The Report of the Iraq Inquiry'*, *House of Commons Report* HC264, 2016.

76 Ministry of Defence, *The Defence and Security Public Contracts Regulations*, 1 August 2013, Chapter 9.
77 Joseph N Nye Jr, 'East Asian Security: The Case for Deep Engagement', *Foreign Affairs*, July/August 1995.
78 BBC, 'London 2012: 13,500 troops to provide Olympic security', *BBC News*, 15 December 2011, www.bbc.co.uk/news/uk-16195861.

Bibliography

De Araujo, JFFE (2001), 'Improving Public Service Delivery: The Crossroads between NPM and Traditional Bureaucracy', *Public Administration*, 79:4, pp. 915–32.

Babcock International (2014), 'Babcock selected as preferred bidder for the acquisition of Defence Support Group', 19 November 2014, www.babcockinternational.com/en/News/Babcock%20selected%20as%20preferred%20bidder%20for%20the%20acquisition%20of%20Defence%20Support%20Group, accessed 18 June 2015.

Babidge, S, Greer, S, Henry, R and Pam, C (2007), 'Management Speak: Indigenous Knowledge and Bureaucratic Engagement', *Social Analysis*, 51:3, pp. 148–64.

BBC (2018), 'London 2012: 13,500 troops to provide Olympic security', 15 December 2011, www.bbc.co.uk/news/uk-16195861, accessed 13 January 2018.

BBC News (2016), 'MoD "wastes money storing unnecessary supplies"', *BBC News*, 28 June 2012, www.bbc.co.uk/news/uk-18614372, accessed 11 July 2016.

Bishop, P (2003), 'Competition and Collaboration in the Provision of Public Services: The Case of the UK Defence Sector', *Journal of Finance and Management in Public Service*, 3:1, pp. 13–24.

Brighton, A (2002), Introduction – 'Management Speak: a master discourse', *Critical Quarterly*, 44:3.

Broadbent, E (1988), 'Military and Government: From Macmillan to Heseltine', *RUSI Defence Studies Series*, 1988, pp. 59–83.

Chilcott, J (2016), *'The Report of the Iraq Inquiry', House of Commons Report* HC264.

Codner, M and Clarke, M (2011), *A Question of Security: The British Defence Review in an Age of Austerity* (London: IB Tauris).

Cowper-Coles, S (2008), 'From Defence to Security: British Policy in Transition', *Survival*, 36:1, pp. 142–61.

Craven, WF, Cate, JL, Ferguson, AB, Fagg, JE, Angell, JW, Goldberg, A, George, RH, Simpson, AF, Finney, RT, Warren, H, Rempel, DG and Goldman, MR (1983), *The Army Air Forces in World War II, Office of Air Force History*, Washington DC.

Defence Infrastructure Organisation (2014), 'MOD unveils facilities management contract', *News story*, 19 March 2014, www.gov.uk/government/news/mod-unveils-facilities-management-contract, accessed 12 October 2016.

Directorate-General for External Policies of the Union, 'The Role of Private Security Companies (PCSs), 2011' in CSDP *Missions and Operations*, European Parliament, EXPO/B/SEDE/FWC/2009-01/LOT6/10/REV1.

Dodd, T (1994), *Frontline First: The Defence Costs Study*, Research Paper 94/101.

Doorman, A (2002), *Defence under Thatcher* (Basingstoke: Palgrave Macmillan).

Dorman, A, Smith, M and Uttley, M (eds, 2002), *The Changing Face of Military Power: Joint Warfare in an Expeditionary Era* (Basingstoke: Palgrave Macmillan).

Edmonds, M (1998), 'Defence Management and the impact of Jointery', *Defense Analysis*, 14:1, pp. 9–27.

Ekström, T and Dorn, M (2014), 'Public Private Partnerships in Defence Acquisition', in Eβig and Glas (eds), *Performance Based Logistics*, Springer, pp. 303–24.

Gady, F (2017), 'India Clears Purchase of 6 US-Made Attack Helicopters', *The Diplomat*, 18 August 2017, https://thediplomat.com/2017/08/india-clear-purchase-of-6-us-made-attack-helicopters/, accessed 11 December 2017.

Gillibrand, A and Hilton, B (1998), 'Resource Accounting and Budgeting: Principles, Concepts and Practice—The MoD Case', *Public Money and Management*, 18:2, pp. 21–8.

Greatbatch, D and Clark, T (2005), *Management Speak: Why we listen to what Management Gurus tell us* (Oxon: Psychology Press).

Heidenkamp, H (2012), 'Sustaining the UK's Defence Effort: Contractor Support to Operations Market Dynamics', *RUSI Whitehall Report* 2–12.

Heidenkamp, H, Louth, J and Taylor, T (2013), *The Defence Industrial Triptych: Government as customer, sponsor and regulator*, RUSI Whitehall Paper 81.

HM Government (2010), *Securing Britain in an Age of Uncertainty: The Strategic Defence and Security Review*, CM7948 (London: The Stationary Office).

HM Government (2015), *National Security Strategy and Strategic Defence and Security Review 2015: A Secure and Prosperous United Kingdom*, CM9161 (London: The Stationery Office).

Hockley, CJ, Smith, JC and Lacey, LY (2011), *Contracting for availability and capability in the defence environment, Complex Engineering Service Systems* (Springer), pp. 237–56.

House of Commons Committee of Public Accounts (2008), *The Privatisation of QinetiQ*, Twenty-Fourth Report of Session 2007–08, HC 151.

House of Commons Defence Committee (2015), *Decision-Making in Defence Policy*, Eleventh Report of the Session 2014–15.

Hoyle, C (2017), 'Thales lands support contract for Watchkeeper UAV', *FlightGlobal*, 22 March 2010, www.flightglobal.com/news/articles/thales-lands-support-contract-for-watchkeeper-uav-339740/, accessed 12 December 2017.

International Security Information Service (1991), *Options for change: the UK defence review, 1990–91*.

Johnman L (1996), 'The Privatisation of British Shipbuilders', *International Journal of Maritime History*, 8:2, pp. 1–31.

Kaplan, RS and Norton, DP (1992), 'The balanced scorecard: Measures that drive performance', *Harvard Business Review*, January–February 1992, pp. 71–9.

Kaplan, RS and Norton, DP (1996), 'The Balanced Scorecard: Translating Strategy Into Action', *Harvard Business Press*.

KBR (2016), 'KBR selected by UK Ministry of Defence for Operational Support Capability Contract', 10 May 2016, https://kbr.com/about/newsroom/press-releases/2016/05/10/kbr-selected-by-uk-ministry-of-defence-for-operational-support-capability-contract.

KBR (2017), 'KBR Awarded Contract to Deliver Facilities Management Services in the Middle East for UK MOD', *KBR press release*, 9 October 2017, www.kbr.com/about/newsroom/press-releases/2017/10/09/kbr-awarded-contract-to-deliver-facilities-management-services-in-the-middle-east-for-uk-mod, accessed 15 December 2017.

Kinsey, C (2009), *Private Contractors and the Reconstruction of Iraq: Transforming Military Logistics* (London: Routledge).

Louth, J (2010), *A Low Dishonest Decade: Smart Acquisition and Defence Procurement into the New Millennium* (Cardiff: University of Wales Institute).

Louth, J and Boden, R (2014), 'Winging It? Defence Procurement as Risk Management', *Financial Accountability & Management*, 30:3, pp. 303–21.

McInnes, C (1998), 'Labour's Strategic Defence Review', *International Affairs*, 74:4, pp. 823–45.

Ministry of Defence (1981), *The UK Defence Programme: The way forward*, CM8288.

Ministry of Defence (1983), *Value for Money in Defence Equipment Procurement*, Defence Open Government Document 83/01.

Ministry of Defence (1998), *Strategic Defence Review: Modern Forces for the Modern World* (London: The Stationary Office).

Ministry of Defence (2001), *British Defence Doctrine*, JWP 0-01.

Ministry of Defence (2002), *Building an Air Manoeuvre Capability: The Introduction of the Apache helicopter*, HC1246, 31 October 2002.

Ministry of Defence (2011), *Defence Reform: An independent report into the structure and management of the Ministry of Defence* (London: The Stationary Office).

Ministry of Defence (2013), *The Defence and Security Public Contracts Regulations*, 1 August 2013.

Ministry of Defence (2014), *Strategic Trends Programme: Global Strategic Trends – Out to 2045*, Fifth Edition, www.gov.uk/government/publications/global-strategic-trends-out-to-2045, accessed 24 April 2015.

Ministry of Defence (2014), *Appraisal Report: Ministry of Defence and the Armed Forces 1963–2014*, 31 July 2014.

Ministry of Defence (2014), *Defence Joint Operating Concept, Joint Concept Note* 1 14.

Ministry of Defence (2015), *Personnel Support for Joint Operations*, Joint Doctrine Publication 1–05, www.gov.uk/government/uploads/system/uploads/attachment_data/file/469880/20151014-Pers_spt_JDP_1_05_Secured.pdf.

Ministry of Defence (2015), 'MOD Operational Support Capability Contract', 13 November 2015, www.government-online.net/mod-operational-support-capability-contract/.

Ministry of Defence (2017), '£1.1 billion investment by MOD in future military flying training', *News story*, 2 February 2016, www.gov.uk/government/news/11-billion-investment-by-mod-in-future-military-flying-training, accessed 11 October 2017.

Ministry of Defence (2017), 'Defence Secretary announces £55m contract for UK bomb disposal robots at DSEI', *News story*, 13 September 2017, www.gov.uk/government/news/defence-secretary-announces-55m-contract-for-uk-bomb-disposal-robots-at-dsei, accessed 11 December 2017.

Ministry of Defence (2018), *Ministry of Defence Annual Report and Accounts 2016 to 2017*, HC 21 2017–2018.

Ministry of Defence, 'Partnership: taking the Whole Force Approach in defence', *Defence Contracts Online*, www.contracts.mod.uk/do-features-and-articles/partnership-taking-the-whole-force-approach-in-defence/, accessed 15 November 2017.

National Audit Office (1987), *Ministry of Defence: Sale of Royal Ordnance Plc*, HC 352 Session 1987–1988.

National Audit Office (2004), *Ministry of Defence: Major Projects Report 2003*, HC 195 Session 2003–2004.

National Audit Office (2006), *Ministry of Defence: Delivering digital tactical communications through the Bowman CIP programme*, HC 1050 Session 2005–2006.

National Audit Office (2008), *Ministry of Defence: Major Projects Report 2008: Project Summary Sheets*.

Norton-Taylor, R (2011), 'MPs attack MoD over £8bn weapons project waste', *Guardian*, 22 February 2011, www.theguardian.com/politics/2011/feb/22/mod-weapons-project-8bn-waste, accessed August 2015.

Nye Jr, JN (1995), 'East Asian Security: The Case for Deep Engagement', *Foreign Affairs*, July/August 1995, www.foreignaffairs.com/articles/asia/1995-07-01/east-asian-security-case-deep-engagement, accessed 14 January 2018.

O'Neill, S (2017), 'Desperate navy seeks senior servicemen', *The Times*, 13 February 2017, www.thetimes.co.uk/article/desperate-navy-seeks-senior-servicemen-5tznx96zf, accessed 11 July 2017.

Parker, D (2009), *The Official History of Privatisation Vol. I: The Formative Years 1970–1987* (London: Routledge).

Peters, HM, Schwatze, M and Kapp, L (2017), *Department of Defense Contractor and Troop Levels in Iraq and Afghanistan: 2007–2017*, Congressional Research Service, 28 April 2017.

Sweetman, J (1984), 'Crucial Months for Survival: The Royal Air Force, 1918–19', *Journal of Contemporary History*, 19:3, pp. 529–47.

Taylor, C (2003), *UK Defence Procurement Policy*, Research Paper 03/78, pp. 16–46.

Taylor, T (2003), 'Jointery: Military Integration', in Teri McConville and Richard Holmes (eds), *Defence Management in Uncertain Times* (Psychology Press), pp. 70–89.

Think Defence (2015), 'MoD Private Finance Initiatives', *Think Defence*, 21 January 2015, www.thinkdefence.co.uk/2015/01/mod-private-finance-initiatives/, accessed 14 June 2017.

Thompson, G (2017), *For Effect or Affect? UK Defence Change: Management*, PhD Thesis, University of Roehampton, 2017.

Twort, L (2015), *Peace and Recovery: Witnessing Lived Experience in Sierra Leone*, PhD Thesis, Business School University of Roehampton, London.

Uttley, M (2005), *Contractors on Deployed Military Operations: United Kingdom Policy and Doctrine*, Strategic Studies Institute.

Warner, J (2016), 'The MoD is still losing the war on waste', *Telegraph*, 9 June 2016, www.telegraph.co.uk/news/2016/06/09/the-mod-is-still-losing-the-war-on-waste/, accessed 11 September 2016.

4 Defence as technology

Introduction

As we have discussed in earlier chapters, the United Kingdom constructs its notions of defence and security within the context of formal international Western alliances, such as NATO, and less formal coalitions such as those engaged at various times in Iraq from 1991 and Afghanistan from 2001. The UK also sees itself as an advanced industrial state with modern military forces. Because of these two factors, defence has, at its heart, a key assumption of military superiority over known and potential state adversaries with this superiority rooted in greater technological innovation and application. This, in turn, is a function of research and development, both sponsored by government and sometimes undertaken within government but more typically within the commercial sector.

Technology, its availability, security and renewal, is central to advanced military states. The design, construction, safe operation, and continued maintenance of a nuclear-powered submarine, for example, is one of the outstanding tests of scientific and engineering capability. Mankind was not supposed to live and work in a canister under the ocean next to an active nuclear reactor surrounded by intercontinental ballistic missiles, nuclear warheads, and other dangerous and toxic materials. Yet this is exactly the scenario witnessed by the use of Polaris, Trident, and its future strategic deterrence submarine force. Whatever personal views are held of nuclear weapons and the arguments relating to deterrence, it is an amazing technological achievement.

Yet the West's confidence in its defence technological superiority which, we contest, has been a feature since 1945, is now being tested by the confluence of emerging, confident powers such as China and the ubiquity and availability of complex technologies. This potential erosion in technological advantage across some capabilities is hugely significant, as this chapter will attest.

Chapter objectives

By the end of this chapter the reader will understand:

1 The role of science and technology within the UK defence system;
2 The challenges posed by the potential closing of the technology gap between Western forces and emerging/challenging powers;
3 The governmental and managerial challenges generated by the need to secure and harness current and future technologies to the defence effort;
4 The types of technologies that could disrupt the current concepts of military operations;
5 The perceived importance of technological innovation to the military instrument.

Chapter structure

This chapter starts with a consideration of the historical role of technology in defence since the end of the Second World War. It goes on to discuss the changes in roles and commitments in research and development funding in government and the private sector and the importance of science and technology to the defence effort. The authors then consider the types of technologies that are set to disrupt defence before discussing the centrality of innovation to the defence effort.

Research and development in government and commerce

People in defence frequently use phrases such as research and development or science and technology, often using them interchangeably. This is a mistake and can lead to confusion. Some definitions and consistency would be helpful.

Definitions of research and development

Research and development is both an experimenting function and a core accountancy practice. Organisations that wish to maintain or improve their technological advantage – for profit maximisation for a business, or to keep us safe and secure from the perspective of government – often have to spend material amounts of money and effort on research into new technologies or new applications for existing technologies. There are two types of research:

• *Pure Research:* Which is work directed purely towards the advancement of knowledge, most typically through expenditure on science and technology enquiry.
• *Applied Research:* Which is work targeted towards the exploitation of pure research towards a specific, identified aim or outcome.

Additionally, there is *development expenditure*, which is the use of scientific or technical knowledge in order to produce new or substantially improved products, materials or practices.

When we talk about science and technology expenditure in defence, we are typically referring to pure research, or research not directed to a specific, identified or measurable outcome aligned to a particular defence programme. As we shall come on to discuss below, in the UK, defence science and technology expenditure is set at 1.2 per cent of the total defence budget, or about £0.4 billion. Applied research and development expenditure can be found across the Equipment Plan and is in the billions.

Technology and defence

Given these definitions as background, we can assert that Western military effort is a highly technical endeavour and has been since the industrial revolution. Historically, as part of the alliance that defeated Nazism and Japanese imperialism during the Second World War, through NATO and via a supposed special relationship with the United States, our military has depended significantly on overmatch of capabilities derived through advances in science and technology. The First Gulf War, in 1991, for example, demonstrated the raw power of military technology, with the Western coalition destroying Iraqi defences through the overwhelming use of technology-rich airpower and land forces. Scientific enquiry, modern industrial design and production processes, smart engineering, and bright ideas are at the core of modern military prowess. It cannot be overstated how much the application of science and technology has underpinned our concepts of defence and security at least since the end of the Second World War.

Yet technology in the twenty-first century is fast moving and the process of incorporating the latest technologies into Britain's military platforms in a cost-effective and timely manner is far from straight forward. So, any insight we glean from a modern historical view must be tempered by the sense that the lessons from yesterday, in relation to defence technologies, may not seamlessly apply to today or tomorrow:

> In an environment where rapid change is a fact of life, our current capability development paradigm is inadequate. Large, complex programmes with industrial era development cycles measured in decades may become obsolete before they reach full-rate production.[1]

Defence technology since 1945

If people ever wonder about technology in the context of the British military they tend to think of two or three distinct things: first, that ships, aircraft, and tanks look highly technical and advanced; second, that those ships, aircraft, and tanks, or their earlier variants, helped the UK to win the Second World War; and, third, that those very ships, aircraft, and tanks so obviously belong to the state that the technologies inside them must belong also to the state. The picture is more complicated than this suggests.[2]

Following the Second World War, for 50 years the globe was divided into two armed camps. NATO, led politically and militarily by the United States, faced off against the Warsaw Pact, with the Soviet Union at its fulcrum. The bipolar world of the Cold War posited that the strategic threat posed from the Soviet Union called for a very high rate of military modernisation, re-capitalisation and capability development in the Western powers. This would lead to strong financial support for defence businesses from governments and strong order books against which investment in business development and equipment prototypes could play out. Specifically in the United States, which was the engine of this defence re-capitalisation in the West, this was underscored by government spending on research, development, testing, and evaluation, both in-house and with the private sector.

This same trend was mirrored in the UK until the Soviet Union collapsed in 1991. For the ten years following this, the post-Cold War honeymoon up to the strategic shock of the 9/11 attacks saw a period of cuts in defence spending chip away at governmental investment in science and technology. Moreover, as budgets contracted:

> There has been a shift towards the cost of full-scale weapons development and away from long-term research. This shift has been driven by an enormous increase in the complexity of the weapons systems that continuously expand the state of the art in each area.[3]

The shift away from the long-term – research in science and technology – to the short-term – development linked to a procurement programme – has been observable. Sir Peter Luff, as the government minister responsible for defence equipment, support, science, and technology observed to the authors, this trend was far from welcomed.

> When I was Minister in 2011 I was prepared to fall on my sword over the question of a ring-fenced, protected budget for science and technology research. I managed to get a commitment that 1.2 percent of the defence budget would be spent on science and technology, but I was fought tooth and nail all the way by my ministerial colleagues and, sadly, some senior people in the military who should have known better.[4]

This is significant in the context of defence spending as a whole. UK defence expenditure has steadily decreased from the historic level of approximately 7 per cent of gross domestic product (GDP) in 1955–1956, to about 4 per cent in 1990–1991, at the end of the Cold War. As Figure 4.1 indicates, from 1969 until 1988 (the year before the fall of the Berlin Wall) the UK spent between 4 per cent and 5 per cent of GDP on defence annually. This was substantially more than all of the UK's NATO allies, with the exception of the United States.

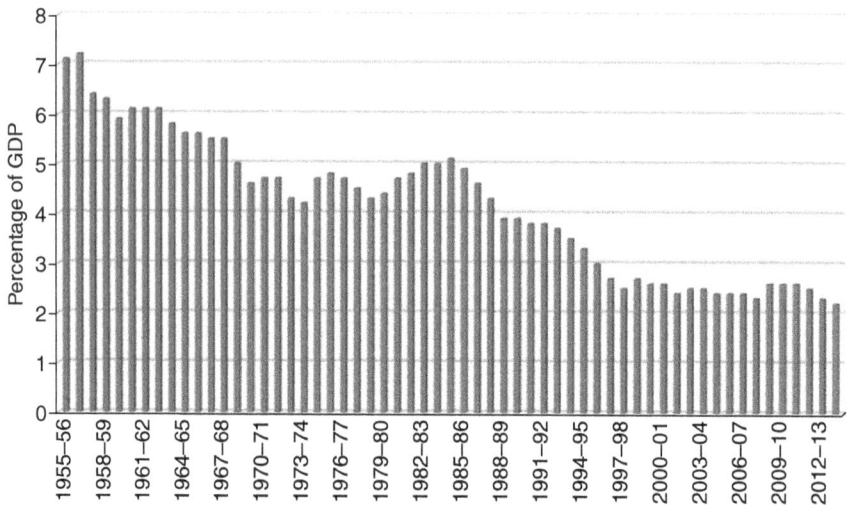

Figure 4.1 UK defence spending as percentage of GDP.

Source: authors' review of National Audit Office Annual Reports to the House of Commons Defence Select Committee.

The Strategic Defence and Security Review (SDSR) in 2010 resulted in a reduction of close to 8 per cent in defence spending. This has led to a 20 per cent reduction in the UK's conventional military combat capability. In 2013, with UK GDP at £1.61 trillion and a defence budget of £37.1 billion, defence expenditure totalled 2.3 per cent of GDP. This is in contrast to UK GDP in 2014 at £1.7 trillion and a defence budget of £36.9 billion, resulting in expenditure of 2.17 per cent of GDP; a reduction in real terms from the previous year of 5.7 per cent.

This has impacted upon science and technology spending. Expenditure on science, engineering, and technology was £1.7 billion in 2015, a drop of £41.8 million, or 2 per cent, since 2014. More strikingly, in constant prices since 2004, defence expenditure has decreased by £1.0 billion, or 39 per cent. Whilst defence research fell by 23 per cent over this period to £0.6 billion, defence experimental development fell by 45 per cent to £1.1 billion. In contrast, over the same period, whilst defence expenditure on science and technology contracted so dramatically, other departments of government were increasing their expenditure on science and technology, from £2.2 billion per annum in 2004 to £3.1 billion in 2015.

UK defence expenditure and technology: today and tomorrow

The current NATO recommended expenditure of at least 2 per cent of GDP on defence was conceived in 2006 to address a perceived imbalance between

American, British, and European NATO contributions, and reaffirmed at the September 2014 NATO Wales summit. In the financial statement of 8 July 2015, the UK Chancellor George Osborne stated that the government is 'committing today to meet the NATO pledge to spend 2 per cent of our national income on defence' per year up to 2020: a commitment that saw defence join the ranks of health, schools, and international development as a 'protected' department with a ring-fenced budget. This supposedly included an annual real-terms increase of half of 1 per cent until 2020–2021.

Accompanying this pledge, the chancellor also announced the creation of a Joint Security Fund (JSF), offering up to an additional £1.5 billion per annum. Additional military and intelligence allocation from the JSF has been estimated to contribute an increase of 1 per cent per annum in real terms, with exact contributions to defence spending being subject to successful bids by the Ministry of Defence to the Treasury for funding from this source. This, of course, is more than the UK spends on defence science and technology.

Of interest, the commitment to spend 2 per cent of GDP on defence is facilitated by significant alterations to the criteria used to calculate the UK defence budget that is reported to NATO. Using the criteria applied prior to the July 2015 budget announcement, the predicted UK defence expenditure for 2015–2016 was £36.8 billion, equivalent to 1.97 per cent of GDP; short of the 2 per cent target. By contrast, the revised criteria resulted in a £39 billion spend in 2015–2016, equivalent to 2.08 per cent of GDP. Manipulating the criteria used to define defence expenditure has therefore enabled the UK to add around £2.2 billion to its NATO-reportable figure. The revised criteria incorporate items previously not included by the UK in defence expenditure calculations: war pensions (around £820 million), civilian MoD pensions (around £200 million), contributions to UN peacekeeping missions (around £400 million), and a large proportion of MoD income. Such dancing on the head of a pin ignores the real managerial need to invest in the key technologies that should bring advantage and, if necessary, victory over peer states.

The technological challenge for defence

The scientific and technological challenge for the UK in the sphere of defence is rooted in today's geopolitical situation and the realities of economic and social globalisation. The Cold War supposed certainties of defence, deterrence, and a rules-based international system, where state actors somehow knew how to play the game of international politics, are giving way to disruption and transformative change, with this caused, in part, by the rate of technological transformation itself. As a consequence, defence policy is no longer just concerned with the military instrument – the armed forces and their supply chains – but rather we need to consider the security and safety of our critical national infrastructure. Hospitals, power stations and the energy grid, financial services, transport hubs, and even supermarkets are susceptible to technological disruption with this, potentially, having catastrophic consequences for society. Food, water, and the security of supply will become increasingly relevant to defence and security policy making.

There is the internet of things that frame our lives and identities, and profoundly powerful communication systems and information technologies seem now to be indispensable to society and citizen alike. The problem for the defence practitioner is that this dependency and all-pervasive network brings with it acute, and new, vulnerabilities that need defending.

The manner in which the UK flexibly adopts and adapts technologies in this search for security has become a defence imperative. We can only begin to imagine the ways in which biomedicine, artificial intelligence, robotics, the internet, and quantum computing will transform the UK and its place in the world. Our ability to be safe and prosperous, in part at least, is dependent upon how we can protect, secure, and assure these strategic modern utilities, or how quickly we can recover from a shock – the resilience test. As Lord Arbuthnot, the ex-defence minister, states: It would be foolish to think that these technologies will not change our constructs of defence and notions of a secure society.[5] The technical challenge for UK defence today is profound, perhaps greater than at any time in British military history. On the one hand, defence equipment is expensive to design, test, and manufacture, can be in service for decades, can be costly to operate, and must form part of an effective and relevant force that can defeat Britain's enemies. On the other hand, technological development is so rapid and new technologies potentially so disruptive to our existing forces and equipment that, what might be relevant and world-leading today, might become highly irrelevant and easily interdicted tomorrow. As General Sir Christopher Deverell, speaking as Chief of Joint Operations, said in 2017:

> How does the politician and commander deal with the cavalry to tank question – typically a once in an epoch step-change in technology on the battlefield making an old technology redundant or less relevant, in this case the horse – when the pace of technological change could pose that question weekly?[6]

There is no quick answer to this challenge of course but it is, perhaps, the military question for the age.

Disruptive technologies

Central to this dilemma is the existence and continuing development and generation of a whole range of technologies and applications that disrupt, or potentially interdict, existing UK (and allies) defence and security platforms and networks, not to mention our critical national infrastructure. For state actors and, increasingly, terrorist organisations are generating or have access to technologies and techniques that threaten UK platforms on the sea, in the air, and on land. This means that Britain's ability to deploy military force, either alone or within an alliance, to strategically significant areas of the world – the Baltic, Gulf, or, perhaps, the South China Sea – is increasingly contested and the risks associated with deployment rise. In addition, these potential adversaries continue to explore disruptive technologies that threaten the UK's use of space for surveillance, communication, navigation, and the like. Offensive cyber capabilities are also being developed to disrupt or diminish our critical national infrastructure on shore. As Dr Andrew Tyler, Chief Executive of Northrop Grumman Europe, says:

> It would not be accurate to assert that potential adversaries have 'caught up' across the full range of defence capabilities, but they have effectively focused their efforts, particularly on sensors, space denial, many extended-range precision missiles able to attack targets on land, sea and in the air, as well as cyber. There are thus growing challenges for UK forces, especially those concerned with force projection.[7]

Given this, it would be useful to explore one or two examples of disruptive technologies in defence. We will discuss directed energy weapons, such as lasers, and unmanned combat air vehicles.

Directed energy weapons

For many years now, proponents of directed energy weapons have claimed 'game-changing' capabilities.[8] Within the United States, for example, the Defense Advanced Research Projects Agency (DARPA) has funded numerous directed energy development programmes such as the Advanced Tactical Laser (ATL), the Area Defense Anti-Munitions (ADAM) ground-based laser, and the Active Denial System (ADS) non-lethal microwave projector. In December 2014, it was announced that the first deployment of a military-grade laser on a commissioned military vessel had taken place. Testing had been undertaken aboard the USS *Ponce* in the form of a technology demonstrator weapon.[9] This advanced stage of development indicated that the US Navy was planning to move forward with fielding this technology operationally in the years ahead.[10] Such technologies will attract other nations and we can expect to see matching Chinese programmes soon.

Laser weapons systems in their current form have shown the defensive capability to engage incoming indirect fire projectiles such as mortar and rocket rounds, as well as being able to 'spoof' the seeker head on incoming infrared (IR) missiles such as the SA-7 Strela-2. Defensive 'spoofing' laser systems are already being deployed on advanced US, Israeli and Russian armoured vehicles, as well as commercial airliners in parts of the world where the proliferation of IR-seeking man-portable air-defence systems (MANPADS), especially among terrorist groups, is a danger.[11]

The future operating environment beyond 2020 is therefore likely to see this sort of technology employed on a wide variety of military platforms in the air, land, and maritime domains. This will have the effect of reducing the vulnerability of installations and some air platforms to low-end missile and indirect fire threats. It may also boost the ability of non-stealth air platforms to penetrate the airspace of sub-peer opponents. However, these defensive laser systems are unlikely to significantly change the current offensive–defensive military equation in the near future since many common threats such as semi-active radar homing missiles, anti-radiation missiles, and gunfire cannot be countered by lasers. Specifically, in the land and maritime domains, defensive laser systems are unlikely to be capable of full-spectrum protection in the near future since ballistic and hypersonic threats offer very limited laser engagement times and are by necessity capable of withstanding extreme heat due to the speeds at which they travel. These threats are, therefore, resistant in their terminal phases to destruction by heat transfer – that is, by laser. This does not, however, rule out the possibility that radical innovation in beam control algorithms, fibre-optic beam technology, capacitors, or some other related field may significantly change this projected trend.

In contrast, offensive directed energy systems certainly have the potential to change significantly the military balance if current limitations on range and energy transfer can be overcome. Though constrained by the requirement for line-of-sight targeting and the curvature of the Earth, as well as being vulnerable to inefficiency in inclement conditions, offensive laser weapons systems offer the potential to bypass conventional defensive systems such as missiles and radar-directed gun systems such as the Phalanx. They also offer potentially unlimited ammunition and the benefits of low cost per shot if sufficient power generation and capacitors are available on the firing platform. However, current-generation laser systems remain vulnerable to beam distortion by moisture or dust at long range in anything other than completely clear weather and therefore require huge amounts of power generation to achieve destructive effects on distant targets that rival low-end conventional munitions. This entails either very large generators or capacitor banks for solid-state lasers or fuel for chemical lasers which is, by definition, limited.[12] In the next ten to 15 years, such systems may well enter military service but, due to these issues, will almost certainly remain a complementary niche capability alongside more traditional systems.

Alongside laser technologies, the Innovative Naval Prototype (INP) Railgun contracts awarded to BAE Systems and General Atomics for deployable systems to be sea-trialled by the US Navy in 2016 represent another significant development in this regard. By using a vast amount of electrical power and sophisticated electromagnetic arrays, these weapons accelerate a streamlined metal slug to hypersonic velocities of above Mach 7, which gives them over 30 MJ of kinetic energy on impact.[13] By comparison, the most up-to-date 120 mm armour-piercing, fin-stabilised, discarding-sabot (APFSDS) tank gun rounds used by most NATO main battle tanks, firing depleted uranium or tungsten alloy penetrators, can achieve around 10 MJ at point blank range.[14] This formidable kinetic energy yield means that railgun projectiles do not require explosive fillers and are therefore smaller, cheaper, and safer than traditional munitions.[15] As with offensive laser systems, the deployment of such weapons by the United States, and later by competing nations, may make defending aircraft, ships, and installations from attack significantly more problematic. If the claims of railgun effectiveness in the defensive role against ground, sea, and air platforms can be realised, platforms armed with such systems will also be significantly harder to attack for any opponent.[16]

Given the costs of fielding such a system from scratch, the United States is likely to be the only power to field innovative systems until the technology is fairly mature, whereupon competitor nations will swiftly move to catch up. However, in this instance it is worth noting W Brian Arthur's observation that 'elaborations, cumbersome as they are, allow the mature technology [in this case missiles] to perform better than its nascent rivals. These may offer future potential, but in their infant state do not perform as well. They cannot directly compete.'[17]

At present, the US Navy is not attempting to incorporate railgun technology into a full-scale procurement programme as the technological risk and full operational utilities are not fully quantified, let alone proven. Instead, it is running parallel-technology, demonstrator-style development projects through BAE and General Atomics.[18] This keeps cost and risk far lower than attempting to develop such novel technology within a full-scale procurement programme, which would entail adapting unproven technology to the shifting demands of serial production, armed-service end users, and changing mission requirements. At the same time, the US Navy has taken care to ensure that the latest warship designs such as the *Gerald R Ford* aircraft carriers and *Zumwalt* destroyers have significant excess power-generation capabilities, which would allow them to mount railgun and directed energy systems if and when the technology is judged mature. In consequence, the US Navy stands to achieve an indeterminate but finite time period as sole operator of functional railgun technology during the next ten to 15 years and, with that, a significant operational advantage within that window.

Given that many of these systems are currently approaching field trials on combat platforms and at a practical scale for attaining initial operating capability (IOC) within the next decade, it seems fair to assume that they will be a part of the future operating environment. They certainly offer the possibility of rapidly improving offensive and defensive capabilities for the United States in particular in conventional warfare terms. However, whilst only the United States is at the stage where this technology's use on a large scale could become a reality in the near future, once it is deployed it is certain that other countries will start to develop competing systems of their own. Furthermore, looking back at many examples of 'game-changing' technology introduced by the United States – such as precision-guided cruise missiles, infrared homing air-to-air missiles, and nuclear and thermonuclear weapons – it is clear that, in each case, it took substantially less time and funding for competing nations to field their own versions of such weapons than the United States had expended in developing the technology in the first place. This trend is likely to accelerate in the future given the difficulty of keeping any large-scale, cutting-edge programme secure from cyber-espionage, as was shown by the 2013 cyber-theft of critical design and performance data on the F-35, Patriot PAC-3, Terminal High Altitude Area Defense (THAAD) anti-ballistic missile system, and Aegis programmes.[19]

Unmanned combat air vehicles

Rapid progress in the field of RPAS and UAVs is another known Science and Technology trend which is likely to continue, ultimately having a significant effect on the military instrument within the future operating environment. However, the prediction often promoted by the media that soon all systems will be unmanned should be treated with some scepticism. The drawn-out counter-insurgency operations in Iraq and Afghanistan have prompted the United States to invest heavily in UAV technology that had been under development in many parts of the world since the 1970s. Reliable satellite datalinks, lightweight con-struction techniques, significant advances in lightweight sensor suites, and precision-guided weaponry have finally enabled RPAS to take a central role in the projection of Western airpower on the battlefield. Platforms such as the MQ-1 Predator and the MQ-9 Reaper offer extreme endurance on station (over 30 hours with upgrades), allowing persistent battlefield surveillance, extensive pre-strike target assessment, and battle damage assessment.[20] They are also much cheaper to operate than their manned strike aircraft alternatives, whilst carrying a precision weapons payload over great distances. The great success of these intelligence, surveillance and reconnaissance (ISR) and so-called 'hunter-killer drones' in Afghanistan in particular have led to predictions by politicians and the media that the future of air combat is unmanned.

It is important to note, however, that there are significant drawbacks to such RPAS platforms in relation to their manned counterparts. Most obviously, they cannot offer the response time that fast jets are famous for. If an RPAS is not already on station when ground forces come under fire or a high-value target is located, one can only transit to the target area at a maximum speed of 300 mph (while many fast jets can achieve over 1,000 mph). Of course, advanced RPAS platforms could be developed to match this performance but they would then lose most of the endurance and cost-saving advantages that appear to be the main attraction of unmanned systems today.

The long-endurance, 'hunter-killer' Predator and Reaper platforms also cannot be operated in a contested air environment as they lack the situational awareness, defensive systems, performance, and agility to defend themselves adequately against either air- or ground-based threats. Future UCAVs designed for operations in defended airspace are likely to closely resemble current stealth UCAV technology demonstrators such as BAE Systems' Taranis and Dassault/ Saab's nEUROn. Like these demonstrators, future UCAVs will most likely be stealthy, supersonic, and compact, with internal weapons carriage and a long range in relation to their size. However, the datalinks that are used to control current RPAS platforms from remote ground stations are vulnerable to disrup-tion or jamming. Satellite uplinks are easy for a moderately advanced adversary to jam, intercept, or 'spoof'. This may well have been the fate of the American RQ-170 Sentinel stealth UAV captured by Iran in December 2011.[21] If a UCAV possessed advanced stealth features and defensive aids, it would be much easier for an opponent to simply interrupt or attempt to override a control uplink rather than targeting the platform physically. This would at minimum result in the UCAV being unable to complete its mission and could result in the loss of the platform.

The likely answer to this problem is UCAVs with significant or total operational autonomy. BAE Systems' Taranis and possibly the European nEUROn technology demonstrators already possess significant autonomous capabilities. They point the way towards a future in which autonomous systems are tasked with carrying out missions which they then complete without further human input. However, this is a significant area for debate as an autonomous weapon release is politically, ethically, and legally highly controversial.[22]

Furthermore, there are also many current uses of combat aircraft that it is difficult to envisage a UCAV or RPAS performing. Quick reaction alert (QRA) is an excellent example. QRA involves either intercepting and escorting foreign military aircraft near national airspace, whilst gathering visual intelligence on any changes in tactics or equipment, or intercepting civilian airliners that are either unresponsive or in trouble. In the last resort, a QRA interceptor could be required to shoot down a hijacked airliner if it were flying towards a built-up or high-value area and no other options were available. Such an act would require a direct order from the head of government and it may be difficult to conceive of any politician authorising such use of lethal force, in the homeland especially, by a 'drone' with the associated controversy surrounding 'remote' or even 'autonomous' use of lethal force.

Optionally piloted systems, which can be flown in both a manned and unmanned configuration, are another potential area of development. There have been several statements to the effect that the US Air Force's (USAF) Long Range Strike Bomber (LRS-B) is being designed to be 'optionally manned'.[23] However, this design factor is unlikely to become particularly common in the foreseeable future since it greatly increases system complexity and therefore cost without really delivering the main advantages normally associated with either configuration. Having to include all of the life-support, information, and safety equipment, as well as the physical space required for aircrew, means that the range, size, and weight savings usually associated with unmanned vehicles are lost. At the same time, including the capability to operate in an unmanned configuration entails greatly increased software and hardware complexity, datalinks, and increased vulnerability to cyber-attack. If significant operational autonomy is built into the system to escape reliance on vulnerable datalinks then serious questions arise from an ethical and legal standpoint, especially if the platform is intended to fulfil a nuclear-capable role as is the case with the LRS-B. As Lieutenant General (ret.) David Deptula, who was responsible for policy formulation, planning, and leadership of USAF ISR and remotely piloted aircraft, observed in 2011, 'There is absolutely no reason, rationale or policy that would support flying remotely piloted aircraft with nuclear weapons onboard.'[24]

What is more likely in the future is advanced semi-autonomous UCAVs operating as 'wingmen' for manned combat aircraft in a networked unit. This would increase the tactical flexibility and weapons payload of a manned platform without building large numbers of expensive fighter aircraft. This would also offer an advantage in terms of increasing the attrition tolerance of air forces which currently possess high-capability, low-mass fleets.

There are, of course, myriad new and emerging technologies way beyond the topics of directed energy weapons, their derivatives, and unmanned combat air vehicles. These examples suggest, merely, that defence is exposed to change generated by new technologies and rapid development. If the decision-makers and military leaders fail to take heed of this they will be left behind. As General Sir Nicholas Carter, Chief of the General Staff, notes:

> I have always to have my eyes on the future. This means I have to build a force that is sleek, adaptive, open to new ways of working and technological insertion. With a military rooted in tradition, training the soldier to think differently and embrace modernity is the millennial challenge.[25]

Innovation

The development of these disruptive technologies and their utility within the hands of the UK's state-peer adversaries as well as existing and new terrorist groups suggests that significant defence innovation – new threats, and new ways of working – is probable as we transit through the twenty-first century. The UK government has responded to this potential change by investing in its own defence innovation initiative, committing £800 million of additional funds over a decade for research.[26] This is in addition to the assured 1.2 per cent of the defence budget that is established for science and technological pure research. Given this, it can be deduced that the core principles of innovation for the UK MoD are:[27]

- A broad and systematic approach that seeks to embed innovation throughout the MoD's organisation, workforce, process, and culture and includes better integration of military concepts, emerging technologies, and capability development.
- A culture that is innovative by instinct by incentivising and rewarding innovative behaviours that we value. Such a culture emphasises the willingness – indeed expectation – to accept risk responsibly across the enterprise.
- An open innovation ecosystem that capitalises on innovative expertise at the MoD and other national security departments and builds effective, efficient, and fertile partnerships with innovators in industry and academia, as well as with key allies and partners.
- The ability to accelerate promising innovations from idea to solution quickly and affordably.
- A strategy-driven approach that provides clear strategic direction to the MoD, the component parts of which will remain primarily responsible for delivering innovation.

As a set of guiding principles these seem sensible but, for us, innovation should not be viewed merely as a management intervention or policy imperative, but a considered, committed, and permanent state of being. It is cultural at its being with the nature of innovation invariably disruptive and messy. An organisation can set the conditions for innovation but it can never 'deliver' it on demand.[28]

Innovation within ways of working and across science, technology, and engineering is the key condition of the UK being able to meet the range of challenges posed by this century. These were characterised by the Secretary of State for Defence, Sir Michael Fallon, as being:[29]

- Project military power against sophisticated adversaries;
- Influence potential adversary choices on terms favourable to the UK;
- Deliver non-traditional and novel ways to have effect beyond traditional weapons systems against sophisticated adversaries;
- Understand and take effective decisions in the information age;
- Adapt with agility to anticipated changes in the strategic environment;
- Maintain robust strategic deterrence into the future;
- Optimise the future workforce to meet anticipated needs.

As a consequence, the procurement arm of the MoD, the Defence Equipment and Support organisation, has developed its Innovation Strategy, highlighting the critical role that body has to play in bringing innovative and niche technologies into the defence order of battle quickly and effectively. Critically, the MoD recognises that this is not simply a matter of horizon scanning and procurement. Rather, the core issue is this ability for Defence as an organisational entity to change its own ways of doing business in order to become battle-winning. Innovation is the perfect storm of behaviours, attitudes, ways of working, and new technologies. The message seems to be that, if the UK defence enterprise gets this wrong, the celebrants will be peer adversaries and terrorist groups.

However, whilst part of the MoD clearly values this imperative to innovate, many key official documents treat the subject sparingly if not glibly:

- The MoD's Global Strategic Trends document covering the period to 2045 has forecasts about the development of laser, directed energy, and even biologically targeted weapons, but there is little sense that an era of rapid innovation may be underway or needed.[30]
- The official How Defence Works guide mentions innovation once but in the context of process rather than technological innovation: it refers to the aim of 'a culture of innovation and efficiency, removing needless process and flushing out bureaucracy.'[31]
- UK Defence Doctrine discusses innovation on just three of its 70-plus pages, the first of which treats the conceptual dimension as a solved problem, arguing that 'the conceptual component is also updated by conceptual innovation, capturing how our thinking changes over time in response

to new technologies, structures and challenges.'[32] The second reference, in the context of the stressed centrality of the adaptability of the human being, simply asserts that 'commanders should also champion innovative practice'[33] while the third, in order to support dealing with the unexpected, makes clear that UK forces should be able to 'adopt the latest technology and systems'. The overall tone of BDD is that people are normally more important than technology.

Whilst we acknowledge that innovation is messy, it seems reasonable to assert that the UK government's policy ambition should at least be joined-up. If innovation is the method for delivering relevant, mission critical technologies and practices as the country faces an uncertain future, as a theme it should feature across all of the MoD's publications and policy guidance.

There is also a broader problem. If innovation is the core principle suggested there is a need to recognise within government and, no doubt, industry as well, that innovation invariably generates a range of losers as misdirected scarce resources as well as winners and smart returns on investment. The change innovation engenders disrupts existing organisations and people and can often signify that the things we hold dear are less relevant than they once were. Such disruption is one of the reasons why history shows many cases of militaries, including that in the UK, being reluctant to adapt to new realities or even the existence of different possibilities.[34]

Currently, legislation is in place that dictates that military force levels should not fall below certain arbitrary numbers. However, this limits the manner in which commanders can balance their spending on people in uniform, those drawn from the contractor community, and the equipment and support bought. Significantly, it also limits the pool of funds available for new research or for development investment – the central planks of an innovation strategy. Should the size of the armed force be freed up, the MoD would be liberated to innovate across the whole of the defence enterprise, but obviously some long-standing organisations would find themselves losing out.

It is significant that government policy is for the UK to have generated a 'full spectrum' force by 2025.[35] This means that the UK will possess a full range of maritime, land, air, space, and cyber capabilities to operate alone or in partner with allies across all threat spectrums. This is intellectually constipated: those threat spectrums will evolve quickly and in unexpected directions just as others benefit from the speed and comprehensiveness of technological development – the UK will always be chasing to stand still if it clings to a full-spectrum ambition. Instead, innovation could – perhaps, should – involve selection, specialisation, adoption, and adaptability. Perhaps it involves deliberate and explicit dependence on allies for some functions, as they, in turn, rely on the UK. The very point of innovation is that what is, so-called, 'full spectrum' today may well change by 2025 and will definitely have a different meaning by 2040. To believe otherwise is to assume that innovation, conceptually, lacks the power we ascribe to it. We should also note that:

Innovation also usually involves a good deal of experimentation and failure, with the body of literature seeming to suggest that organisations should aim to experiment a lot, diagnose prospects at an early stage, abandon the failing before too much money has been spent, and always learn lessons. This involves a greater readiness to write off public money than appears to be present yet in UK political culture. Policy disclosure of a government readiness to accept the risks of an emphasis on innovation would be a prior condition for changed attitudes across the defence enterprise and broader society, and especially perhaps within the House of Commons Public Accounts Committee.[36]

In the modern epoch, businesses and non-governmental organisations have found it difficult to move from established methods of working and existing products, despite profound changes in our markets. The existence of so many change management consultancies and a multi-billion-dollar global organisational transformation industry points to this fact.[37] Typically, businesses do not require their employees to put their lives at risk. How much more difficult it must be for the armed forces and MoD more generally to embrace change and a constant demand for innovation. Consequently, the challenges of championing, embedding, and utilising a defence innovation culture should not be underestimated, perhaps not least because the armed forces rely so significantly on tradition and appreciation of a common history for the generation and maturation of motivation and commitment at both the unit and individual levels. But it is a challenge that the MoD cannot ignore.

Conclusion

As the discussion in the chapter has demonstrated, technology of all types is central to notions of defence in the UK. Indeed, the West practices defence through the acquisition and maintenance of technological advantage – if that is eroded by the ambitions, policies, and practices of peer and near-peer rival states, a core building block of the security of the UK could be undermined. The criticality of programmes of research and development, both within government and the commercial sector sponsored by government, cannot be downplayed and remains a topic of political controversy within Whitehall and Westminster, as the topic goes to the heart of defence budgeting, planning, and strategic expenditure. This is especially so in a period of high modernity, such as now, where so many technologies emerge that are truly innovative and disruptive of the old norms. Defence is hardly immune to 'progress' in this regard.

Notes

1 US Air Force (USAF), 'America's Air Force; A Call to the Future,' July 2014, p. 10, http://airman.dodlive.mil/files/2014/07/AF_30_Year_Strategy_2.pdf, accessed 11 March 2015.

2　See Jacques S Gansler, *Democracy's Arsenal: Creating a Twenty-First-Century Defense Industry* (Cambridge MA: MIT Press, 2011) p. 9.

3　Ibid., p. 27.

4　Speaking to the authors in February 2015 and again in May 2017.

5　John Louth, Trevor Taylor and Andrew Tyler, *Defence Innovation and the UK: Responding to the Risks Identified by the US Third Offset Strategy, RUSI Occasional Paper*, 2017, p. V, Foreword.

6　In conversation with the authors at the British Army Land Warfare Conference in London, Tuesday 27 June 2017.

7　Speaking to the authors in July 2017 ahead of a joint publication on defence innovation.

8　Ian Bellany and Coit D Blacker, *Antiballistic Missile Defence in the 1980s* (London: Frank Cass and Company Ltd, 1983), p. 42.

9　America's Navy, 'Historic Leap: Navy Shipboard Laser Operates in Arabian Gulf', 10 December 2014, www.navy.mil/submit/display.asp?story_id=84805, accessed on 5 March 2015.

10　The Boeing/General Atomics electromagnetic railgun system is also of interest in this context. Whilst it is not a directed energy weapon, it also operates without chemical propellants or explosives, requires the same scale of capacitor and power-supply improvements to platforms, and offers to extend the range and firepower of ships whilst greatly reducing projectile cost.

11　Tamir Eshel, '"Skyshield" – Counter MANPADS Laser Countermeasure Completes Testing', *Defence-Update.com*, 27 February 2014, http://defense-update.com/2014 0227_skyshield-dircm-test.html#.U7pg2ECwWSp, accessed 11 March 2015.

12　A solid-state laser is a laser that uses a gain medium that is a solid, rather than a liquid such as in dye lasers or a gas as in gas lasers. In practical terms, this means that military lasers working on this principle draw on electrical power stored in capacitors to generate a beam, rather than chemical lasers which use a chemical fuel to do so. The advantage of this is that with adequate power generation capabilities, 'ammunition' is effectively unlimited.

13　Office of Naval Research, 'Electromagnetic Railgun', 2014, www.onr.navy.mil/media-center/fact-sheets/electromagnetic-railgun.aspx, accessed 11 March 2015.

14　Orbital ATK, '120 mm Ammunition', 2014, www.orbitalatk.com/products-services/120mm-ammunition, accessed 11 March 2015.

15　Allen McDuffee, 'Navy's New Railgun Can Hurl a Shell Over 5,000 MPH', *Wired.com*, 9 April 2014, www.wired.com/2014/04/electromagnetic-railgun-launcher/, accessed on 19 May 2014.

16　General Atomics, 'Railgun Systems', *Press release*, 2015, http://media.ga.com/video-library/railgun-capabilities/, accessed 11 March 2015.

17　W Brian Arthur, *The Nature of Technology: What is it and How it Evolves* (London: Penguin Books, 2009), pp. 138–39.

18　BAE Systems, 'Electromagnetic (EM) Railgun', 2015, www.baesystems.com/product/BAES_158879/electromagnetic-em-railgun, accessed 11 March 2015; see also General Atomics, 'Railgun Systems', 2015, www.ga.com/railgun-systems, accessed 11 March 2015.

19　Ellen Nakashima, 'Confidential Report lists U.S. Weapons System Designs Compromised by Chinese Cyberspies', *Washington Post*, 27 May 2013.

20　General Atomics, 'GA-ASI Unveils New Enhanced Endurance Designs for Predator B', *Press release*, 18 April 2012, www.ga-asi.com/news_events/index.php?read=1&id=388&date=2012, accessed 11 March 2015.

21　Dave Majumdar, 'Iran's Captured RQ-170: How Bad Is the Damage?', *Defense News.com*, 9 December 2011, www.defensenews.com/article/20111209/DEFSECT01/112 090307/Iran-s-Captured-RQ-170-How-Bad-Damage, accessed 11 March 2015.

22 See Birmingham Policy Commission, 'The Security Impact of Drones: Challenges and Opportunities for the UK', University of Birmingham, October 2014.

23 Dave Majumdar, 'U.S. Air Force May Buy 175 Bombers', *Defense News.com*, 23 January 2011, www.defensenews.com/article/20110123/DEFFEAT04/101230303/U-S-Air-Force-May-Buy-175-Bombers, accessed 11 March 2015.

24 Ibid.

25 Speaking to the authors at the Military Academy, Sandhurst on 29 October 2015.

26 Ministry of Defence, *Advantage Through Innovation: The Defence Innovation Initiative* (London: MoD, 2016).

27 Ibid., p. 4.

28 See Martin C Libicki, *Cyberspace in Peace and War* (Annapolis: NIP, 2016).

29 Speaking at the British Army Land Warfare Conference in London, Wednesday 28 June 2017.

30 Ministry of Defence, *Strategic Trends Programme: Global Strategic Trends – Out to 2045*, 5th edition, pp. 94–6.

31 Ministry of Defence, *How Defence Works*, November 2015 edition, p. 6, www.gov.uk/government/uploads/system/uploads/attachment_data/file/484941/20151208How DefenceWorksV4_2.pdf, accessed 12 January 2016.

32 Ministry of Defence, *UK Defence Doctrine*, The Development, Concepts and Doctrine Centre, November 2014, p. 3.

33 Ibid., p. 38.

34 Terry Pierce, *Warfighting and Disruptive Technologies: Disguising Innovation* (Oxon: Taylor & Francis, 2004).

35 Ministry of Defence, 'SDSR 2015 Defence Fact Sheets', p. 1.

36 John Louth, Trevor Taylor and Andrew Tyler (2017), op. cit., p. 18.

37 See John Louth, *Expert or Charlatan? The Rise and Rise of Management Consulting* (New Delhi: KW Publishing, 2014).

Bibliography

America's Navy, 'Historic Leap: Navy Shipboard Laser Operates in Arabian Gulf', 10 December 2014, www.navy.mil/submit/display.asp?story_id=84805, accessed on 5 March 2015.

Arthur, WB (2009), *The Nature of Technology: What is it and How it Evolves* (London: Penguin Books).

BAE Systems, 'Electromagnetic (EM) Railgun', 2015, www.baesystems.com/product/BAES_158879/electromagnetic-em-railgun, accessed 11 March 2015.

Bellany, I and Blacker, CD (1983), *Antiballistic Missile Defence in the 1980s* (London: Frank Cass and Company Ltd).

Birmingham Policy Commission, 'The Security Impact of Drones: Challenges and Opportunities for the UK', University of Birmingham, October 2014.

Eshel, T (2014), '"Skyshield" – Counter MANPADS Laser Countermeasure Completes Testing', *Defence-Update.com*, 27 February 2014, http://defense-update.com/2014 0227_skyshield-dircm-test.html#.U7pg2ECwWSp, accessed 11 March 2015.

Fallon, M (2017), Speech at the British Army Land Warfare Conference in London, 28 June 2017.

Gansler, JS (2011), *Democracy's Arsenal: Creating a Twenty-First-Century Defense Industry* (Cambridge MA: MIT Press).

General Atomics, 'Railgun Systems', *Press release*, 2015, www.ga.com/railgun-systems, accessed 11 March 2015.

General Atomics, 'GA-ASI Unveils New Enhanced Endurance Designs for Predator B', *Press release*, 18 April 2012, www.ga-asi.com/news_events/index.php?read=1&id=388&date=2012, accessed 11 March 2015.

Libicki, MC (2016), *Cyberspace in Peace and War* (Annapolis: NIP).

Louth, J (2014), *Expert or Charlatan? The Rise and Rise of Management Consulting* (New Delhi: KW Publishing).

Louth, J, Taylor, T and Tyler, A (2017), *Defence Innovation and the UK: Responding to the Risks Identified by the US Third Offset Strategy, RUSI Occasional Paper*.

Majumdar, D (2011), 'Iran's Captured RQ-170: How Bad Is the Damage?', *Defense News.com*, 9 December 2011, www.defensenews.com/article/20111209/DEFSECT01/112090307/Iran-s-Captured-RQ-170-How-Bad-Damage, accessed 11 March 2015.

Majumdar, D (2011), 'U.S. Air Force May Buy 175 Bombers', *Defense News.com*, 23 January 2011, www.defensenews.com/article/20110123/DEFFEAT04/101230303/U-S-Air-Force-May-Buy-175-Bombers, accessed 11 March 2015.

McDuffee, A (2014), 'Navy's New Railgun Can Hurl a Shell Over 5,000 MPH', *Wired.com*, 9 April 2014, www.wired.com/2014/04/electromagnetic-railgun-launcher/, accessed on 19 May 2014.

Ministry of Defence (2014), *UK Defence Doctrine*, The Development, Concepts and Doctrine Centre, November.

Ministry of Defence (2014), *Strategic Trends Programme: Global Strategic Trends – Out to 2045*, 5th edition.

Ministry of Defence (2015), 'SDSR 2015 Defence Fact Sheets', www.gov.uk/government/uploads/system/uploads/attachment_data/file/492800/20150118-SDSR_Factsheets_1_to_17_ver_13.pdf, accessed 1 February 2018.

Ministry of Defence (2015), *How Defence Works*, November 2015, www.gov.uk/government/uploads/system/uploads/attachment_data/file/484941/20151208HowDefenceWorksV4_2.pdf, accessed 12 January 2016.

Ministry of Defence (2016), *Advantage Through Innovation: The Defence Innovation Initiative* (London: MoD).

Nakashima, E (2013), 'Confidential Report lists U.S. Weapons System Designs Compromised by Chinese Cyberspies', *Washington Post*, 27 May 2013.

Office of Naval Research (2014), 'Electromagnetic Railgun', www.onr.navy.mil/media-center/fact-sheets/electromagnetic-railgun.aspx, accessed 11 March 2015.

Orbital ATK (2014), '120 mm Ammunition', www.orbitalatk.com/products-services/120mm-ammunition, accessed 11 March 2015.

Pierce, T (2004), *Warfighting and Disruptive Technologies: Disguising Innovation* (Oxon: Taylor & Francis).

US Air Force (USAF) (2014), 'America's Air Force; A Call to the Future,' July 2014, http://airman.dodlive.mil/files/2014/07/AF_30_Year_Strategy_2.pdf, accessed 11 March 2015.

5 Defence as industrial policy

Introduction

The attitudes of the British governmental defence community towards the private sector have evolved significantly over centuries and continue to do so. Defence industrial policy, therefore, is a complicated and increasingly challenging subject, as this chapter makes clear.

Chapter objectives

By the end of this chapter the reader will understand:

1 The government's increasing reliance on the private sector for the delivery of defence capabilities;
2 The strategic significance of an on-shore, national defence supply base;
3 The significance of economic considerations when deciding upon a defence industrial policy and strategy;
4 The development and purposes of a national defence industrial policy;
5 The role of the nuclear deterrence;
6 The place of shipbuilding in the UK defence industrial policy.

Chapter structure

This chapter starts with a review of the UK's increasing dependence on commercial firms and practices, taking a distinctly historical perspective. We then go on to consider the on-shore supply base and its importance to defence before considering broader economic and employment considerations. The final sections deal with the strategic deterrence, nuclear power systems, and the national shipbuilding strategy.

Towards increased reliance on the private sector

In the earlier days of the development of professional armed forces, there was common reliance to establish state bodies to develop and furnish specialised military equipment, not least because the private sector was not so developed as to be able to raise the capital needed to finance such activities. State armouries (which eventually became the Royal Ordnance group of facilities) supplied much army equipment while the Navy designed its own ships and built them in its own dockyards. For instance, this was the case with the remarkable Dreadnought fleet, whose demanding requirements were specified by the Navy in 1905, with the first ship being delivered in 1907.

However, as defence platforms grew more complicated and the links grew to civil products (as with the ties between armoured vehicles and the car and truck sector, and between the civil and military aerospace sector), and as the whole of the economy was mobilised for war between 1914 and 1918 and between 1939 and 1945, the norm became the *Private Manufacture of Armaments* (to use the title of Philip Noel-Baker's critical book).[1]

The period after 1945 was marked by a significant reduction in demand levels, which was moderated for a period in Britain by the demands of the Korean War. However, there was also requirement for ever-more complicated platforms incorporating extensive technological innovation. These were associated with significant cost increases and, over time, with smaller fleet sizes and longer intervals between orders. The inevitable consequence was the consolidation of platform and major subsystem suppliers. Under the Labour governments of the 1970s, the adopted solution involved nationalisation with the creation of British Aerospace (BAe) and British Shipbuilders in 1977 to sit alongside the already state-owned Royal Ordnance and Royal Dockyards, as well as the Atomic Weapons Establishment (AWE) responsible for the development and production of British nuclear warheads and associated systems.

The Thatcher governments after 1979 believed that private sector enterprises were inherently more efficient than state bodies and better able to raise capital.[2] There was also a need to raise funds for government, which opted over time to sell off nationalised industries across the board, including in the defence sector.[3] BAe (like British Telecom) was floated in tranches on the stock market, whereas British Shipbuilders was broken up and yards sold to four individual buyers by 1988. Royal Ordnance was sold to British Aerospace in 1987, and the Naval Dockyards were either closed or completely in private hands by 2011. AWE was made into a government-owned contractor-operated body[4] in 1993, a status it continues to enjoy. The Ministry of Defence's research institutions that had been bundled into the Defence Evaluation and Research Agency were split up, with most capabilities being passed to a new company called QinetiQ that was privatised in two main stages from 2003. The last step to date in the privatisation trend was the sale to Babcock in 2015 of the Defence Support Group, which was primarily concerned with the repair and upgrade of armoured vehicles.[5]

However, the growth in weight of the private sector in defence was not just a matter of the privatisation of state-owned firms and agencies. Two key ideas prompted the MoD to entrust ever more of its work to the private sector. The first was the noted idea, particularly associated with New Public Management thinking, that state bodies were almost invariably less efficient and agile than private-sector bodies.[6] The second was the more generic management concept of core competence most associated with CK Prahalad and G Hamel.[7] This asserted that no organisation could be very good at everything it needed to have done, and therefore it should focus its internal efforts on the selected fields where it needed to and could excel. For the other things it required, it should rely on external suppliers who were focused and excellent. In government terms, the MoD saw a range of areas, not least in facilities management, which were not seen as fields of military core competence.

These factors drove the MoD to spend more of its money with the private sector rather than to do things internally and thus raised the importance of the MoD's capacity to deal with the private sector. By 2016 the MoD was spending about £24 billion of its £37 billion budget with the private sector.

That private sector was increasingly owned by companies based outside the UK, with the growth of Thales, Leonardo, and Airbus, as well as US firms including Lockheed-Martin, Raytheon, General Dynamics, and Northrop-Grumman. Explicitly since the Defence Industrial Policy of 2001,[8] the government has defined the UK branches of these companies as British firms, and indeed the vast majority of their employees are British citizens. What is less clear, however, is how these firms would behave in terms of supporting British forces if the UK was to launch an operation that was not supported by the government where the company's headquarters was located.[9]

The perceived strategic significance of the defence supply base

A useful perspective on the MoD is to view it as the prime contractor responsible for the generation of UK military capability, capability to be at the disposal for use by government when circumstances require action. Prime contractors in the commercial world, for instance those producing cars, computers, and so, pay great attention to the performance and health of the supply chains and networks on which they depend for production, improvement, and their reputation. The Japanese car firm Toyota is known in part for its demanding but close and supportive attitudes towards its suppliers: it is a significant element of 'the Toyota way'.[10] A fundamental message emerging from a focus on supply chain is that what happens at the production end of that chain depends on what goes on from the very beginning.[11] However, constancy of concern about suppliers has been a feature of British government behaviour for decades.

British defence policy and practices have not reflected a consistent or coherent stance towards defence industry in the UK.[12] To offer a simplified but reasonable summary of policies, through the 1980s the government promoted competition in defence procurement. If that competition could not be British because of the presence of just one supplier, it was opened to international bidders, most notably from the United States. Under the Thatcherite notion that the government should not be responsible for directing the direction of economic development, there was no industrial policy and thus no defence industrial policy was articulated or pursued. But, in 1985, when it appeared that Westland Helicopters might be taken over by Sikorsksy of the United States, the government split seriously on the issue, with it being apparent that little prior thought had been given to the strategic significance of different elements of defence industrial capability. Ministers had to make up their responses on the day, as it were, and famously the dispute led to the resignation as Defence Minister of Michael Heseltine.[13] The key decision confirming an increased readiness to buy from abroad was the cancellation of the AEW Nimrod programme in 1987 and the purchase instead of a fleet of E3A AWACs aircraft from Boeing. From the time of Peter (now Lord) Levene as part of the UK government, the only criterion for maintaining a UK industrial capacity in a specific area was if it was impossible to buy a technology from overseas because of legal or political restrictions.

An important element of context for the exposure of UK defence industry to external competition was that, for 45 years after the end of the Second World War, the increasingly dominant focus of MoD effort was deterrence of the Soviet Union and its allies. Significantly, neither of NATO's two doctrines (NATO's 'Massive Retaliation'[14] followed by 'Flexible Response'[15] from 1968) contemplated a protracted conflict or the mobilisation of industry. In both doctrines, a readiness to escalate to the nuclear level after a week or so of conventional exchanges was central to deterrence thought. The Third World War would have been fought along planned lines by forces in being. In the Cold War, the industrial base was important for the development and production of new systems for deployment, but not for the support of those systems or for their adaption to fit unforeseen circumstances.

However, for conventional wars an industrial base remained very important. In 1982, the British efforts to re-take the Falklands after their occupation by Argentina required much short-notice effort by British industry to prepare the force as well as some reliance for supplies on a United States that, until quite late in the day, was not certain which side to support in the conflict.[16]

In contrast to the reduced concern with industry during the 1980s, four years after the Labour Party had come to power in 1997, the MoD published its Defence Industrial Policy[17] document, followed in 2005 by an extensive Defence Industrial Strategy (DIS) Paper[18] dealing with the desired UK approach in areas of capability and technology. This was a lengthy and detailed document, specifying that the UK should have the capacity at least to sustain and modify its equipment, even that which had been bought from overseas.[19] It dealt with defence relevant areas of industry on a sector-by-sector basis, specifying levels of ambition for each. It was followed by an equally lengthy Defence Technology Strategy[20] specifying the many fields of science where the UK needed to be working and holding expertise. The then Procurement Minister Lord Drayson was the driver behind much of this work, but the government was never ready to provide the funding that these documents implied. Then the financial crisis that built from 2007 hit the MoD. Lord Drayson resigned and abandoned political life in November 2007.

Jump ahead a further five years and the National Security Through Technology (NSTT) White Paper[21] from the Conservative-Liberal coalition in 2012 asserted that the government's default position would be buy on a competitive basis from the international market unless (unspecified) security considerations directed otherwise.

> We will use competition as our default position and ... we will look at the domestic and global defence and security market for products that are proven, that are reliable, and that meet our current needs.[22]

The Executive Summary reiterated the point:

> Wherever possible, we will seek to fulfil the UK's defence and security requirements through open competition in the domestic and global market, buying off-the-shelf where appropriate.[23]

There was no hint of any preference for UK or European products: an earlier (MoD) draft of the document envisaged as a successor to the 2005 DIS had addressed individual sectors, but this was abandoned on the direction of wider government influences who argued that the UK did not pursue industrial policies.

The Levene years, the 2012 NSTT paper, and indeed elements of the National Shipbuilding Strategy of 2017[24] (see below) reflected a core belief that the best thing that the UK government could do for British defence firms was to subject them to external competition so that they would be driven to pursue efficiency and effectiveness. Underlying this argument was the fear that, if British firms felt assured of government orders, they would become complacent and poorly performing. This must be recognised as a real risk, but the hazards and costs associated with relying massively on external suppliers are also non-trivial.

However, in one notable area, a very different set of behaviours was practised. At the time of the first DIS there was government awareness that its spending on complex weapons (i.e. missiles and their closely associated technologies) was about to decline sharply. The DIS therefore envisaged joint working between the MoD and central relevant companies in the UK led by the multinational joint venture MBDA, so that the UK could maintain a capacity to design, develop, produce, test, and support a range of guided weapons. Its original designation was Team Complex Weapons[25] and the MoD even secured some legal permission from government to privilege the companies involved. Raytheon, a significant employer in the UK and the manufacturer of Paveway II and Paveway IV bombs, declined to be involved because it had some rival products to those of MBDA and also was concerned about US anti-trust legislation.

This arrangement has survived the NSTT and its record shows that favoured firms do not necessarily become complacent. MBDA have led with the successful development of weapons that have been delivered on time and to budget and which perform impressively. The ASRAAM (Advanced Short-Range Air-to-Air

Missile), the Meteor medium range air-to-air missile, the Storm Shadow air-to-ground weapon, the dual-mode Brimstone short-range air-to-ground missile, and most recently the Sea Viper air-defence weapon, all appear as success stories. The agile and extremely accurate dual-mode Brimstone proved invaluable in the Libya campaign of 2011 with its small warhead, and the company itself responded energetically to the MoD's need for surged production during the campaign.[26] Moreover, national partners have supported the maintenance of a European weapons capability, with the French signing up to the FASGW project for a helicopter–launched anti-ship missile being particularly significant.

However, missiles usually have to be integrated into platforms and in the air domain this is usually expensive and sometimes politically difficult if the platform is not controlled by a UK company. The UK has not been able to get Paveway on to its Apache D helicopters, although it is thought a better weapon than the Hellfire it would replace. The Apache E models that the UK intends to buy are meant to be equipped with Brimstone and this appears to have been a condition for the UK purchase. However, other foreign purchasers of Apache will not have the option of equipping them with Brimstone: political and hard national economic considerations are usually present in defence trade matters, including in the United States. In 2017 only the MBDA ASRAAM is hard-scheduled to be integrated on to the F.35 and of course the platform prime contractor always has a loud voice as to what any integration exercise is to cost.

In defence acquisition matters, final victory should never be declared, but the Team Complex Weapons construct has progressed well over more than a decade, and is moving towards the sustainment of a European-supported capability set rather than just a British model. A challenge for governments will be to make sure that the companies have good access to platforms on which to integrate their products.

The orientation of the 2012 NSTT White Paper was reflected in a series of major government procurement choices, except that the formal competition element was sidestepped. The MoD committed to buy a series of US systems basically from American production lines, often using the US Foreign Military Sales (FMS) process.[27] This is an arrangement where purchase from the manufacturing company is done by the US government using a contract, and then the transfer to the actual user government is done on the basis of a government-to-government international agreement. FMS deals formally cannot be won through a formal competition because the US government will not 'bid' its systems against a competing product.

Thus, the UK commitment to US systems grew to include additional Chinook helicopters, C.17 transport aircraft, and Beechcraft Shadow R1 ISTAR aircraft, and new to the UK Rivet Joint electronic warfare aircraft, the F.35 combat aircraft, Reaper/Protector unmanned combat air vehicles, p. 8A maritime patrol aircraft, and Apache E helicopters. Only in the case of the F.35 was there any significant role for UK industry.[28]

Employment and the economy

As UK governments have fluctuated towards and away from a positive approach to the promotion of UK defence industry, the public relations sections of the MoD have seemingly been constant in their determination to highlight the employment consequences of defence choices. MoD press releases about defence contracts always mention jobs creation and sustainment if possible. This was true even for the nine P-8A aircraft that the UK committed to buying from Boeing through the US government in 2016. Thus, an excerpt from the official government announcement of the purchase read as follows:

> The P-8A is based on the Boeing 737, which is already supplied by UK industry, supporting several hundred direct UK jobs. What is more, UK manufacturers also already provide specialist sub-systems for the P-8 itself. Companies include Marshall for the auxiliary fuel tanks, Martin Baker for the crew seats and General Electric for weapon pylons. The new order of P-8As is also set to create opportunities for the UK to bid for training and support contracts.[29]

However, the drafters of press releases seemingly have little influence on procurement decisions. Thus, when the MoD published its first document on Value for Money in Defence Procurement,[30] although there were dozens of variables listed under a number of headings, the impact of choices on the UK economy was not one of them.[31]

In contrast in the United States, employment is a very big consideration in defence acquisition choices, in part because Congressmen and Senators are anxious to build prosperity in their constituencies and have the constitutional power to decide which major projects are to be funded and to what extent. Reflecting this, the United States imports very little defence equipment and, on the rare occasions when foreign systems are bought, they tend to be produced in the United States under technology transfer arrangements: the US Navy trainer aircraft, the Goshawk, which is a variant of the British Hawk trainer, would be an illustration of this. A BAE Systems product, it is built in the United States in a plant owned today by Lockheed Martin.[32]

Back to a defence industrial strategy?

Reflecting that things rarely stand still, the Strategic Defence and Security Review of 2015 indicated that a move back to more widespread explicit support for the UK defence industrial sector was possible.[33]

Consideration of military operations needs to be taken into account. During the protracted campaigns in Iraq and Afghanistan, the MoD had made extensive use of US industry to meet Urgent Operational Requirements, particularly for heavily protected vehicles such as Mastiff.[34] However, there was also a need to mobilise UK industry and then, for the 2011 Libya campaign, a number of British firms responded to the need urgently to modify equipment and to boost production of munitions. Their efforts resulted in a formal appreciation of their contribution from the government, which was reminded of the importance of an assured and agile supply base for the conduct of military operations.

The protracted repercussions of the financial crash dating back to 2008 increased concerns about economic growth and government deficits, and all ministries were given a responsibility for contributing to 'the prosperity agenda'. The promotion of prosperity was Objective 3 in the 2015 SDSR, with detail being spelt out in Chapter 6.[35] Then the prospect of economic damage as a result of the Brexit referendum result in 2016 made the government even more sensitive to take an active role in promoting economic advance, not least in the manufacturing sector. The depreciation of the pound that followed the referendum also significantly damaged the MoD's purchasing power.[36]

Even when the MoD was busy buying off-the-shelf equipment from the United States, small budget but high ambition activities of government cooperation with British industry were continued and expanded. The post-2010 regime continued to support the Niteworks[37] joint MoD-industry organisation which studied potential solutions to defence challenges. It launched first an Aerospace Growth Partnership programme[38] for government-industry discussion and action to promote the aerospace industry overall, and then it set up a Defence Growth Partnership (DGP)[39] to address the sector overall. From the DGP emerged the idea of a Defence Solutions Centre (DSC)[40] as a parallel body to Niteworks and which took over from the Centre for Defence Enterprise. The latter had been a small fund set up under Labour under which small enterprises could bid for money for focused research work without extensive paperwork and being assured of a rapid decision.

Neither Niteworks, the DGP, nor the DSC involved sums of money that could be considered significant in defence terms, and indeed a significant indicator of successive governments' reluctance to focus resource on sustaining the UK defence supply base is the significantly reduced share of the defence budget that has been devoted to research and development since the turn of the millennium and even more markedly since the end of the Cold War.

These points emerge from conversions of cash spending on defence to spending at constant prices (2015).[41] In 1990/1991 British defence spending at 2015 prices was £45.7 billion, which had fallen to £40.6 billion in 2002/2003. It remained broadly at this level, being £39 billion in 2016/2017. As a generalisation, it can be said that defence spending in real terms has been constant since 2002 and has fallen only slightly since the end of the Cold War. Also, in this millennium, the share of the budget devoted to equipment has been broadly constant, fluctuating only between 19.5 and 24.8 per cent of the defence budget. In other words, in real terms, the same amount continued to be spent on equipment.

However, between 2001/2002 and 2014/2015, research spending was reduced by a quarter and development spending more than halved, an implicit but clear consequence of the reliance on foreign products for urgent Operational Requirements during the Iraq and Afghanistan campaigns, and the purchase from American production lines for major air platforms.

The UK government does not publish statistics on its defence imports. However, the US Department of State does offer numbers in its World Military Expenditures and Arms Transfer annual online publication.[42] The figures there are in current dollars, i.e. not adjusted for inflation, but they show clearly a consistent pattern. Although a decline can be noticed as the Afghanistan campaign came to an end, purchases since of F.35s, P-8As, Apache, and Protector, as well as parts and services associated with in-service kit, suggest that future numbers will not fall sufficiently (see Table 5.1).

The clear implication is that British defence industrial strategy has in practice been increasingly focused on three sectors: complex weapons (already discussed), the nuclear 'enterprise', and warships. The latter encompasses the capability to design, build, and support nuclear weapons, nuclear submarines and the reactors that power them, the command and communications with the submarine at sea, and the secure storage and transport of nuclear weapons.

The nuclear enterprise

In order to maintain the long-standing policy commitment that the UK should deploy an independent nuclear deterrent, the MoD has spent significantly to maintain most of the industrial base needed to make this possible.

As well as the development of the bombs themselves, in the 1950s the UK developed and produced three strategic bombers, the Victor, the Valiant and the Vulcan, but backed out of the development of a ballistic missile capability when it cancelled the Blue Streak programme in 1958.[44] It then opted for long-term reliance on a US-origin delivery system first in the form of the planned air-launched Skybolt missile. When that was cancelled by the Americans in 1961, the MacMillan government was able to negotiate the purchase of submarine-launched Polaris missiles.[45] When they needed replacement in the mid-1980s, the government agreed with the United States initially (in 1981) that the UK would buy the newer Trident C4 missile before switching to the emerging, longer-range and more expensive Trident D5 missile. Moreover, instead of buying the missiles and being responsible for their support, an agreement was reached that the UK would lease the weapons, returning weapons for maintenance to Bay in the United States and taking away replacements.[46] Thus the UK abandoned industrial capability with regard to the missile delivery system.

Table 5.1 Imports of defence goods and services in constant 2014 $billion[43]

	2004	2005	2006	2007	2008	2009	2010	2011	2012	2013	2014	Mean
UK	9.6	8.5	9.1	9.3	11.0	11.5	11.6	12.1	12.1	11.6	9.9	10.6
France	2.5	2.8	2.4	1.6	1.7	1.6	1.6	1.6	1.8	1.4	1.1	1.8
Germany	2.8	10.4	5.0	3.0	4.1	3.5	3.6	3.7	3.9	3.0	2.3	4.1
Saudi Arabia	7.0	5.8	5.7	2.3	2.7	5.4	6.3	5.0	5.4	7.9	6.9	5.5

Source: US Department of State, World Military Expenditures & Arms Transfers 2016.

The other elements in the nuclear enterprise, i.e. the weapons, with their targeting, decoys, and re-entry systems, the command and control systems including the firing chain for the weapons, the (delivery platforms) submarines, the nuclear power plants in the boats, and the storage and transport of weapons in the UK, remained a UK responsibility, as did responsibility for the disposal of radioactive material and other components.

Responsibility for the weapons and associated elements, including targeting mechanisms, lies with the government-owned contractor-operated AWE at Aldermaston and Burghfield. It is important to realise that this is an organisation that must deal with research and development, production, testing, and support. Nuclear warheads are machines containing chemicals and other components whose characteristics change with time and so constant development, production, and test work is needed for assurance that they remain both safe and reliable. It involves production activities as weapons need re-assembling.

The UK has a long-standing programme and late-delivery for the development of a new warhead design, although of course the British commitment to the Nuclear Test Ban Treaty means that novel testing techniques are needed. There is a need constantly to review the capacity of the re-entry systems to evade defences and even to improve accuracy. Underpinning all these activities is the requirement to attract and retain very high calibre scientists and engineers by providing them with intriguing and intellectually satisfying work, and providing them with research opportunities is an important element here.

To organise and deliver this work, the government opted to run AWE as a government-owned contractor-operated GoCo facility. The unusual governance mechanism is that AWE Management Limited (AWE ML), which is the vehicle, holds the assets on behalf of the MoD and directs the overall management of the site. It files accounts at Companies House, although its last Annual Review covered 2013.[47] The partner commercial companies in AWE ML can and have been changed, with the last contract (for 25 years) being awarded to a consortium of Lockheed-Martin, Serco, and Jacobs.

Under the direction of AWE ML, AWE plc actually operates plants, employs the people, is responsible for sub-contracting, and overall delivers the outputs. The term plc is a misnomer since it does not file accounts and last published an Annual Review in 2013, most of the AWE plc staff are in fact civil servants working alongside sub-contractors. Plainly the idea is to feed private sector management expertise continuously into a government-dominated body.

Overall MoD figures show that the costs of operating AWE amount to around a billion a year in cash terms, having risen steadily from £311 million since 2003, before stabilising around 2014.[48]

The UK's capability to design, develop, and build nuclear submarines rests centrally on the construction facilities at BAE System's Barrow site,[49] and on its many specialist and often sole source suppliers, including Rolls Royce. Support activities at Faslane are the responsibility of Babcock. Particularly since preparation of the decision to develop the next generation of nuclear bomber submarines (the Successor programme), which was agreed by parliament in 2007, the MoD has pursued a more positive approach to sustaining the industrial capabilities associated with nuclear submarines. However, it had to start from rather a poor base.

Britain's nuclear submarine industrial capacity was much reduced once the programme of Trident boats was completed in 1998. Sustaining the industrial capacity to design and build machines as complex as nuclear submarines requires more than orders every 25 to 30 years for four boats. The run down in employee numbers, but also in skills and knowledge, caused by the gap between orders for the last Trident boat and the first order for the Astute attack submarine is widely recognised as having contributed to the delays, cost increases, and wider problems of the Astute class.[50] Without work for many of its employees, BAE Systems ran down its Barrow workforce from a peak of around 14,000 to some 2,500: in 2006 the Barrow workforce was just over 3,000.[51] As a RAND study for the MoD observed in 2005, 'Force structure reductions and budget constraints have led to long intervals between design efforts for new cases and have not considered industrial base efficiencies resulting in feast or famine for the organisations that support submarine construction'.[52] Recovery, reflected in the delays and cost increases in the Astute programme, was painful although, by 2017, with the Astute programme progressing and the Successor programme approved and in its early stages, employment had risen to 9,000 across submarine business in total, but 5,000-6,000 working on the Successor programme.[53]

A significant specialist element in the nuclear enterprise is the development and production of the reactor that powers the boat. At the time of the Polaris agreement, the UK was given a major boost by the US agreement to transfer relevant technology to Rolls Royce (rumoured to be the only UK British company that Admiral Rickover in the United States trusted to look after the technology).[54] By 2017, Rolls Royce had delivered reactors for a succession of attack and bomber submarines with its established PWR1 and 2 designs.[55] Its current challenge is to design and deliver a new design PWR3[56] to drive the Successor submarines with reduced support needs and improved performance, including noise reduction.

The MoD apparently had to play catch-up with regard to the production infrastructure for the PWR3 when, in 2012, it signed a contract for the modernisation of the reactor core production site at Derby, leading the National Audit to observe:

> 'The Nuclear Propulsion Critical Technology programme brings focused investment to regenerate the UK nuclear propulsion design and support capability, and ensures we have the design base essential to maintain a strategic sovereign nuclear capability.'[57]

This survey covers much but not all of the nuclear enterprise: in particular it omits the provision for the transport and storage of nuclear weapons, and for the communications technologies that enable command and control. These are sensitive matters understood to be undertaken predominantly by government personnel.

Overall, the government's overall approach was to work in close partnership with the key sole source suppliers (BAE Systems, Rolls Royce, and Babcock) in a Submarine Enterprise Performance Programme[58] in order to control costs and deliver the needed capabilities. It demonstrates that, when the pressure is on, the MoD is capable of recognising and implementing supply chain management as a commercial enterprise might.

The UK and building warships

The National Shipbuilding Strategy that the government published in the summer of 2017 recognised the importance of the defence supply base and reflected all the different lines of thought that have characterised UK defence industrial stances.[59] On the one hand, it half recognised a strategic interest of the UK in being able to develop and build its own warships, so there was a commitment to build both the Type 26 and the Type 31 frigates at home.[60] Other relevant factors might be presumed to have been the economic importance of naval shipbuilding on the Clyde and its involvement in reducing the chances of Scottish secession from the UK.

On the other hand, the demands of the Navy were seemingly prioritised over the needs of industry. There was no recognition of the need of business for a steady 'drumbeat' of work or a long-term plan as had been spelled out by RAND studies of surface ships and submarines a decade earlier. Among other points, RAND had said that 'The MoD should attempt to smooth, or 'level load' the production and design demands it places on the industrial base'.[61] The Strategy stressed that the MoD was working on a 30-year master shipbuilding plan for the Navy, and it gave assurance that it would take a portfolio approach to future shipbuilding.[62] However, this plan apparently was not seen as producing anything like constant demand. Instead the Strategy observed that Navy demand would be a matter of 'peaks and troughs'.[63] The MoD's Strategy document was based extensively on a report it had commissioned from Sir John Parker, a report that had not even indicated awareness of the RAND body of work.[64]

The Shipbuilding Strategy clung on to hopes for the feasibility of competition in warship development and construction, and envisaged British shipyards competing for warship work, in part because of the perceived beneficial effects of competition, but also because the Navy was in a hurry to obtain its five Type 31s. Five of these were to be delivered, one annually from 2023, while Type 26 production was to start in 2026, with a delivery every 18 months.[65]

The reliance on competition to secure lower prices was reflected in the stance of the Strategy document that there would be international competitive tendering for the construction of future ships that were not classified as warships: 'Our intent is to compete non-warships in order to maintain UK competitive edge for shipbuilding. By testing UK yards against foreign competition, we will be able to ensure that the UK sector remains competitive'.[66] This was a line of reasoning about defence industry as a whole that had been common in the late 1980s, when the possibilities of state-subsidised external competitors and predatory pricing practices were again not taken explicitly into account.

The expectation in the 2017 Strategy document that exports could be a significant source of construction work reflected significant optimism, since countries that are able to afford frigate-sized ships increasingly want to be able to build/ assemble them in their own yards. As industrialists we have spoken to confirm, British naval exports are most likely in the sub-systems areas, including weapons and engines, rather than in whole ships.[67] The report did not refer to any systemic study of the likely warship market although there was significant reliance on exports to support British firms.

In brief, there was a very different approach to the surface ship-building sector than was the case for nuclear submarines, although the aims in both cases were similar. The National Shipbuilding Strategy observed that 'for reasons of national security, the UK will need to retain the ability to design, build and integrate warships. This industrial capacity enables the UK to sustain the Royal Navy without interference from a foreign power'.[68]

Conclusion

The sentence above encapsulates the importance of the industrial supply base for a country that aspires to be 'strong, influential, global' and which sees the freedom to use national armed forces as its sees fit as reflecting the 'essence of sovereignty'.[69]

However, it also exposes the extent to which many other elements of UK capability, not least the Royal Air Force, is significantly susceptible to interference by a foreign power, not least because all its major surveillance and intelligence platforms are of foreign origin.[70] The mischievous will also ask questions about the aircraft operated on board Navy ships.

In terms of treating defence as being significantly about assured access to a capable supply base, the MoD has not shown consistency of purpose or a comprehensive approach needed at least to control the risks of 'interference by a foreign power'.

Looking forward, in areas where the UK has had significant national capability, including combat aircraft, helicopters, sonar buoys, torpedoes, aircraft defensive aids, tactical communications, and of course armoured vehicles, there are major questions about the sustainment of these capabilities through national and collaborative programmes.

Self-sufficiency in defence industry and technology is clearly unattainable for a country of 60 million, but arguably the reluctance of successive governments to face up to the implications of extensive external dependency and their readiness to cut R&D spending is having serious consequences. In reality, the UK's capacity for undertaking any significant sustained operation depends on a multi-billion pounds a year external supply of spares and advice.

Several middle-sized states face the challenges of striking the right approach and balance of costs and risk between equipping the armed forces with the items they most desire and sustaining a national industrial capability that reduces the negative economic impact of defence spending and which contains external dependence. The UK is not alone, with France, Germany, Japan, and even Australia operating in a similar context. UK governments have fluctuated in their stances and policies, but what definitely should not be overlooked is the very different political context of the period since 1990. The possibilities for sustained operations that come as a surprise to planners are clear. A reliable, capable, and wide ranging extended defence enterprise is vital for dealing with such matters.

Notes

1 Philip Noel-Baker, *The Private Manufacture of Armaments* (New York: Oxford University Press, 1937).
2 See, for instance, ER Yescombe, *Public–Private Partnerships: Principles of Policy and Finance* (Oxford: Butterworth-Heinemann, 2007).
3 Elke Krahmann, *States, Citizens and the Privatisation of Security* (Cambridge: Cambridge University Press, 2010), p. 73.
4 The strategic direction and assets of the company would be decided by the government; however, it would be operated on a for-profit basis by a private company who would be accountable to their shareholders.
5 Ministry of Defence, 'Defence Support Group sold for £140 million' 17 December 2014, www.gov.uk/government/news/140-million-contract-signed-for-sale-of-defence-support-group, accessed 11 September 2017.
6 Tom Christensen and Per Laegreid (eds), *Transcending New Public Management: The Transformation of Public Sector Reforms* (Aldershot: Ashgate, 2007) p. 4.
7 1990.
8 Ministry of Defence, *Defence Industrial Strategy*, Policy Paper No. 5 (London: MoD, 2002).
9 These and other issues are discussed in Henrik Heidenkamp, John Louth and Trevor Taylor, *The Defence Industrial Triptych: Government as Customer, Sponsor, and Regulator*, *RUSI Whitehall Paper* 81 (London: Taylor and Francis, 2013).
10 See, for instance, Jeffrey K Liker, *The Toyota Way: 14 Management Principles from the World's Greatest Manufacturer* (London: McGraw Hill Professional, 2003).
11 See, for instance, Ananth V Iyer, Sridhar Seshadri, Roy Vasher, *Toyota Supply Chain Management: A Strategic Approach to Toyota's Renowned System* (New York: McGraw Hill Professional, 2009).
12 For a fuller exposition of this thinking, see J Louth and T Taylor, 'Beyond the Whole Force: the Concept of the Defence Extended Enterprise', *RUSI Occasional Paper* (London, November 2015), https://rusi.org/publication/occasional-papers/beyond-whole-force-concept-defence-extended-enterprise-and-its, accessed 21 September 2017.
13 BBC, 'On this day: 1986: Heseltine quits over Westland', *BBC*, http://news.bbc.co.uk/onthisday/hi/dates/stories/january/9/newsid_2516000/2516187.stm, accessed 17 October 2017.
14 A strategy of deterrence where an attack from an aggressor would result in the use of nuclear weapons. This became the basis for security for NATO members.
15 Included the use of conventional warfare as well as nuclear weapons.
16 See, for instance, Neil Tweedie and Steven Swinford, 'US wanted to warn Argentina about South Georgia' *Daily Telegraph*, citing opened government archive material,

28 December 2012, www.telegraph.co.uk/news/uknews/defence/9767707/US-wanted-to-warn-Argentina-about-South-Georgia.html, accessed 24 July 2017, accessed 14 November 2017.

17 Ministry of Defence, *Defence Industrial Strategy*, Policy Paper No. 5 (London: MoD, 2002).

18 Ministry of Defence, *Defence Industrial Strategy: Defence White Paper*, CM6697 (London: The Stationary Office, 2005).

19 Ibid., p. 17, 64, 66.

20 Ministry of Defence, *Defence Technology Strategy for the demands of the 21st century* (London: The Stationary Office, 2006).

21 Ministry of Defence, *National Security Through Technology: Technology, Equipment, and Support for UK Defence and Security*, CM8278 (London: The Stationery Office, 2012).

22 Ibid., p. 6.

23 Ibid., p. 8.

24 Ministry of Defence, *The National Shipbuilding Strategy: The Future of Naval Shipbuilding in the UK*, (London: MoD, 6 September 2017).

25 Defined as a value for money approach to delivering the UK's Complex Weapons requirements which also ensures a viable industrial capacity. See official MBDA website www.mbda-systems.com/about-us/mission-strategy/team-complex-weapons/, accessed 10 October 2017).

26 See, for instance, Richard Norton-Taylor, 'Libyan air campaign boosts MBDA order book' *Guardian*, 20 March 2012; Christina Goulter, 'The British Experience: Operation Ellamy', in Karl P Mueller (ed.), *Precision and Purpose: Airpower in the Libyan Civil War* (Santa Monica: RAND, 2015), pp. 153–82.

27 For general information on the purpose, structure and organisation of the FMS process see, US Department of State, 'Foreign Military Sales: Process and Policy', testimony by Tina S Kaidanow, Acting Assistant Secretary, Bureau of Political-Military Affairs, 15 June 2017.

28 See, for instance, George Allison, 'How much of the F-35 is British?' *UK Defence Journal*, 14 July 2016 https://ukdefencejournal.org.uk/british-f-35/, accessed 12 November 2017.

29 Ministry of Defence, 'MOD seals the deal on nine new Maritime Patrol Aircraft to keep UK safe', 11 July 2016, www.gov.uk/government/news/mod-seals-the-deal-on-nine-new-maritime-patrol-aircraft-to-keep-uk-safe, accessed 24 July 2017, accessed 21 August 2017.

30 Ministry of Defence, *Value for Money in Defence Equipment Procurement* (London: The Stationary Office, 1983).

31 Ministry of Defence, *Parker Review: Blueprint for a strong naval shipbuilding sector*, November 2016, www.gov.uk/government/news/parker-review-blueprint-for-a-strong-naval-shipbuilding-sector, accessed 24 September 2017, accessed 14 September 2017.

32 See, for instance, 'First production T-45C Goshawk rolled out at Boeing', *Aviation Week*, 3 November 1997 http://aviationweek.com/awin/first-production-t-45c-goshawk-rolled-out-boeing, accessed 24 July 2017.

33 HM Government, *National Security Strategy and Strategic Defence and Security Review 2015: A Secure and Prosperous United Kingdom*, CM9161 (London: The Stationery Office, 2015), Chapter 6, section C.

34 National Audit Office, *Support to High Intensity Operations* (London: The Stationary Office, 2009).

35 HM Government (2015), CM9161.

36 See, for instance, Trevor Taylor, 'The Ministry of Defence's Post-Brexit Spending Power: Assumptions, Numbers, Calculations and Implications', *RUSI Commentary*, 12 August 2016.

37 For more information see Niteworks official website, www.niteworks.net/, accessed 20 November 2017.

38 See, for instance, John Child, *Organization: Contemporary Principles and Practices*, 2nd edition (Chichester: John Wiley & Sons, 2015), p. 275.

39 See, for instance, Keith Hartley, *The Political Economy of Aerospace Industries: A Key Driver of Growth and International Competitiveness?* (Edward Elgar Publishing, 2014), p. 111.

40 For more information see the official website, www.ukdsc.org/, accessed 14 September 2017.

41 The following data is a digest of the written evidence of Trevor Taylor to the House of Commons Defence committee in January 2017, http://data.parliament.uk/written evidence/committeeevidence.svc/evidencedocument/defence-committee/defence-acquisition-and-procurement/written/44710.html, accessed 22 September 2017.

42 They produce a series of reports on military expenditure and arms transfer, including annual military expenditures, arms transfers, armed forces, economic data, and indicators relevant to the military-economic ratio. For more information see, www.state.gov/t/avc/rls/rpt/wmeat/, accessed 11 October 2017.

43 US Department of State, World Military Expenditures & Arms Transfers 2016, www.state.gov/t/avc/rls/rpt/wmeat/2016/index.htm, accessed 22 September 2016.

44 The Blue Streak programme was a British attempt to manufacture medium-range ballistic missiles to retain the national capability of a nuclear deterrent. The programme was deemed too expensive and too vulnerable, and was cancelled.

45 See, for instance, The National Archives Cabinet Papers on Skybolt and Polaris missiles, www.nationalarchives.gov.uk/cabinetpapers/themes/skybolt-polaris-missiles.htm, accessed 12 July 2017.

46 For more information, see Claire Mills, 'Replacing the UK's "Trident" Nuclear Deterrent', *House of Commons Briefing Paper*, 12 July 2016.

47 AWE, 'Annual Report 2013: Delivering the Deterrent', www.awe.co.uk/app/uploads/2014/07/Annual-Review-2013_FINAL.pdf, accessed 26 September 2017.

48 Nuclear Information Service, *AWE: Britain's Nuclear Weapons Factory: Past, present and Possibilities for the Future*, June 2016, p. 5; AWE, 'Annual report and Financial Statements from the year ended 31 December 2016', P&L account, p. 9.

49 See, for instance, Lauren Twort and Gabriela Thompson, 'The Twenty-First Century Armoury: A Town Called Barrow', *The RUSI Journal*, November 162(5), 2017, pp. 46–73.

50 John F Schank, Frank W Lacroix, Robert Murphy, Cesse Cameron Ip, Mark V Arena and Gordon T Lee, *Learning from Experience: Volume III: Lessons from the United Kingdom* (Santa Monica, CA: RAND Corporation, 2011).

51 House of Commons Defence Committee, 'The Future of the UK's Strategic Deterrent: the Employment & Skills Base', 19 December 2006, para 17, https://publications.parliament.uk/pa/cm200607/cmselect/cmdfence/59/59.pdf, accessed 21 November 2017.

52 John F Schank, Jessie Riposo, John Birkler and James Chiesa, *The United Kingdom's Nuclear Submarine Industrial Base, Volume 1: Sustaining Design and Production Resources* (Santa Monica, CA: RAND Corporation, 2005) www.rand.org/content/dam/rand/pubs/monographs/2007/RAND_MG608.pdf, accessed 27 September 2017.

53 BAE Systems, 'BAE Systems awarded £201m to further Successor Submarine design', Newsroom, 10 February 2016 www.baesystems.com/en/article/bae-systems-awarded-to-further-successor-submarine-design, accessed 15 October 2017.

54 Steve Ludlam, "The Role of Nuclear Submarine Propulsion" in Jenifer Mackby and Paul Cornish (eds), U.S.-UK Nuclear Cooperation After 50 Years, (Washington DC: CSIS, 2008), p. 253.

55 See, for instance, the submarine propulsion section on the official Rolls Royce website www.rolls-royce.com/products-and-services/marine/about-marine/market-sectors/submarines.aspx, accessed 12 September 2017.

56 Ibid.
57 National Audit Office, *Ministry of Defence Major Projects Report 2012* (London: The Stationary Office, 2013), p. 371, www.nao.org.uk/wp-content/uploads/2013/03/Major-Projects-Vol-2.pdf, accessed 27 September 2017.
58 HM Government, *Securing Britain in an Age of Uncertainty: The Strategic Defence and Security Review*, CM7948 (London: The Stationary Office, 2010), p. 39.
59 Ministry of Defence, *National Shipbuilding Strategy: The Future of Naval Shipbuilding in the UK*, 2017, www.gov.uk/government/publications/national-shipbuilding-strategy, accessed 24 September 2017.
60 Ibid., p. 33.
61 RAND Europe, *Naval shipbuilding in the United Kingdom: RAND Europe research summary* (Santa Monica, CA: RAND Corporation, 2007), www.rand.org/content/dam/rand/pubs/research_briefs/2007/RAND_RB9205.pdf, accessed 25 September 2017.
62 Ministry of Defence (2017), op. cit., p. 15.
63 Ibid., p. 39.
64 RAND Europe, *Naval shipbuilding in the United Kingdom: RAND Europe research summary* (Santa Monica, CA: RAND Corporation, 2007).
65 Ministry of Defence (2017), op. cit., p. 24.
66 Ibid.
67 Author interviews.
68 Ministry of Defence (2017), op. cit., p. 33.
69 Ministry of Defence, *National Security Through Technology: Technology, Equipment, and Support for UK Defence and Security*, CM8278 (London: The Stationery Office, 2012), p. 26.
70 This list includes E3A AWACs, Sentinel, P-8A, Protector, Rivet Joint/Air Seeker, Beechcraft King Air Shadow 1 fleet, and Watchkeeper.

Bibliography

Allison, G (2016), 'How much of the F-35 is British?' *UK Defence Journal*, 14 July 2016 https://ukdefencejournal.org.uk/british-f-35/, accessed 12 November 2017.

Aviation Week (2017), 'First production T-45C Goshawk rolled out at Boeing', *Aviation Week*, 3 November 1997, http://aviationweek.com/awin/first-production-t-45c-goshawk-rolled-out-boeing, accessed 24 July 2017.

AWE (2013), 'Annual Report 2013: Delivering the Deterrent', www.awe.co.uk/app/uploads/2014/07/Annual-Review-2013_FINAL.pdf, accessed 26 September 2017.

AWE (2016), 'Annual report and Financial Statements from the year ended 31 December 2016', P&L account.

BAE Systems (2016), 'BAE Systems awarded £201m to further Successor Submarine design', Newsroom, 10 February 2016 www.baesystems.com/en/article/bae-systems-awarded-to-further-successor-submarine-design, accessed 15 October 2017.

BBC, 'On this day: 1986: Heseltine quits over Westland', *BBC*, http://news.bbc.co.uk/onthisday/hi/dates/stories/january/9/newsid_2516000/2516187.stm, accessed 17 October 2017.

Child, J (2015), *Organization: Contemporary Principles and Practices*, 2nd edition (Chichester: John Wiley & Sons).

Goulter, C (2015), 'The British Experience: Operation Ellamy', in Karl P Mueller (ed.), *Precision and Purpose: Airpower in the Libyan Civil War* (Santa Monica: RAND), pp. 153–82.

Hartley, K (2014), *The Political Economy of Aerospace Industries: A Key Driver of Growth and International Competitiveness?* (Edward Elgar Publishing).

Heidenkamp, H, Louth, J and Taylor, T (2013), *The Defence Industrial Triptych: Government as Customer, Sponsor, and Regulator*, RUSI Whitehall Paper 81 (London: Taylor and Francis).

HM Government (2010), *Securing Britain in an Age of Uncertainty: The Strategic Defence and Security Review*, CM7948 (London: The Stationary Office).

HM Government (2015), *National Security Strategy and Strategic Defence and Security Review 2015: A Secure and Prosperous United Kingdom*, CM9161 (London: The Stationery Office).

House of Commons Defence Committee (2006), 'The Future of the UK's Strategic Deterrent: the Employment & Skills Base', 19 December 2006, https://publications. parliament.uk/pa/cm200607/cmselect/cmdfence/59/59.pdf, accessed 21 November 2017.

Iyer, AV, Seshadri, S, and Vasher, R (2009), *Toyota Supply Chain Management: A Strategic Approach to Toyota's Renowned System* (New York: McGraw Hill Professional).

Kendall, F (2013), speech at CSIS, 7 November 2013.

Krahmann, E (2010), *States, Citizens and the Privatisation of Security* (Cambridge: Cambridge University Press).

Liker, JK (2003), *The Toyota Way: 14 Management Principles from the World's Greatest Manufacturer* (London: McGraw Hill Professional).

Louth, J and Taylor, T (2015), 'Beyond the Whole Force: the Concept of the Defence Extended Enterprise', *RUSI Occasional Paper* (London: November 2015), https://rusi. org/publication/occasional-papers/beyond-whole-force-concept-defence-extended-enterprise-and-its, accessed 21 September 2017.

MBDA website www.mbda-systems.com/about-us/mission-strategy/team-complex-weapons/, accessed 10 October 2017.

Mills, C (2016), 'Replacing the UK's 'Trident' Nuclear Deterrent', *House of Commons Briefing Paper*, 12 July 2016.

Ministry of Defence (1983), *Value for Money in Defence Equipment Procurement* (London: The Stationary Office).

Ministry of Defence (2002), *Defence Industrial Strategy*, Policy Paper No. 5 (London: MoD).

Ministry of Defence (2005), *Defence Industrial Strategy: Defence White Paper*, CM6697 (London: The Stationary Office).

Ministry of Defence (2006), *Defence Technology Strategy for the demands of the 21st century* (London: The Stationary Office).

Ministry of Defence (2012), *National Security Through Technology: Technology, Equipment, and Support for UK Defence and Security*, CM8278 (London: The Stationery Office).

Ministry of Defence (2016), 'MOD seals the deal on nine new Maritime Patrol Aircraft to keep UK safe', 11 July 2016, www.gov.uk/government/news/mod-seals-the-deal-on-nine-new-maritime-patrol-aircraft-to-keep-uk-safe, accessed 24 July 2017, accessed 21 August 2017.

Ministry of Defence (2016), *Parker Review: Blueprint for a strong naval shipbuilding sector*, November 2016, www.gov.uk/government/news/parker-review-blueprint-for-a-strong-naval-shipbuilding-sector, accessed 24 September 2017, accessed 14 September 2017.

Ministry of Defence (2014), 'Defence Support Group sold for £140 million', 17 December 2014, www.gov.uk/government/news/140-million-contract-signed-for-sale-of-defence-support-group, accessed 11 September 2017.

Ministry of Defence (2017), *The National Shipbuilding Strategy: The Future of Naval Shipbuilding in the UK*, 6 September 2017.

National Archives Cabinet Papers on Skybolt and Polaris missiles, www.nationalarchives.gov.uk/cabinetpapers/themes/skybolt-polaris-missiles.htm, accessed 12 July 2017.

National Audit Office (2013), *Ministry of Defence Major Projects Report 2012* (London: The Stationary Office), p. 371, www.nao.org.uk/wp-content/uploads/2013/03/Major-Projects-Vol-2.pdf, accessed 27 September 2017.

National Audit Office (2009), *Support to High Intensity Operations* (London: The Stationary Office).

Niteworks official website, www.niteworks.net/, accessed 20 November 2017.

Noel-Baker, P (1937), *The Private Manufacture of Armaments* (New York: Oxford University Press).

Norton-Taylor, R (2012), 'Libyan air campaign boosts MBDA order book', *Guardian*, 20 March 2012.

Nuclear Information Service (2016), *AWE: Britain's Nuclear Weapons Factory: Past, present and Possibilities for the Future*, London: NIS, June 2016.

Prahalad, CK and Hamel, G (1990), 'The Core Competence of the Corporation' *Harvard Business Review*, May–June 1990.

RAND Europe (2007), *Naval shipbuilding in the United Kingdom: RAND Europe research summary* (Santa Monica, CA: RAND Corporation).

Schank, JF, Riposo, J, Birkler, J and Chiesa, J (2005), *The United Kingdom's Nuclear Submarine Industrial Base, Volume 1: Sustaining Design and Production Resources* (Santa Monica, CA: RAND Corporation), www.rand.org/content/dam/rand/pubs/monographs/2007/RAND_MG608.pdf, accessed 27 September 2017.

Schank, JF, Lacroix, FW, Murphy, R, Ip, CC, Arena, MV and Lee, GT (2011), *Learning from Experience: Volume III: Lessons from the United Kingdom's* (Santa Monica, CA: RAND Corporation).

Taylor, T (2016), 'The Ministry of Defence's Post-Brexit Spending Power: Assumptions, Numbers, Calculations and Implications', *RUSI Commentary*, 12 August 2016.

Taylor, T (2017), 'Supplementary written evidence', on Defence Acquisition and Procurement submitted to the Defence Committee, 17 January 2017.

Tweedie, N and Swinford, S (2012), 'US wanted to warn Argentina about South Georgia' *Daily Telegraph*, citing opened government archive material, 28 December 2012, www.telegraph.co.uk/news/uknews/defence/9767707/US-wanted-to-warn-Argentina-about-South-Georgia.html, accessed 24 July 2017, accessed 14 November 2017.

Twort, L and Thompson, G (2017), 'The Twenty-First Century Armoury: A Town Called Barrow', *The RUSI Journal*, November 162(5), pp. 46–73.

US Department of State (2016), 'World Military Expenditures & Arms Transfers 2016', www.state.gov/t/avc/rls/rpt/wmeat/2016/index.htm, accessed 22 September 2016.

US Department of State (2017), 'Foreign Military Sales: Process and Policy', testimony by TS Kaidanow, Acting Assistant Secretary, Bureau of Political-Military Affairs, 15 June 2017.

Yescombe, ER (2007), *Public–Private Partnerships: Principles of Policy and Finance* (Oxford: Butterworth-Heinemann). www.ukdsc.org/, accessed 14 September 2017. http://data.parliament.uk/writtenevidence/committeeevidence.svc/evidencedocument/defence-committee/defence-acquisition-and-procurement/written/44710.html, accessed 22 September 2017.

6   Defence as exports and engagement

Introduction

Defence exports and defence engagement are two increasingly significant elements of UK defence. They are brought together in this single chapter because, although in some ways separate, they are also unavoidably and sometimes uncomfortably related. To a large degree, they have their own separate justifications and drivers, and are implemented through different organisations and personnel. They can both complement each other and be a source of friction, and British defence attaches can find themselves dealing with both.

Chapter objectives

By the end of this chapter the reader will understand:

1   The changing roles and practices of government in relation to defence exports;
2   Defence exports and sales as government activities;
3   The performance of UK exports;
4   Export controls and practices;
5   UK export controls and the development of UK capability requirements.

Chapter structure

This chapter starts with a historical review of defence management for exports and broader engagement. This leads onto exports and sales as inherently governmental activities. We then consider the performance of UK exports and how they relate to capability requirements' generation and the importance, or otherwise, of engaging with and influencing allies through exports.

The changing roles of government

Since 1966 the UK government has, with fluctuating degrees of enthusiasm, supported the export of British defence goods and services, but it should not be forgotten that government involvement at all in the defence trade is, in historical terms, a relatively recent activity. Strange as it seems today, until just before the Second World War, the marketing and sale of military equipment remained activities largely unregulated by governments. When Hiram Maxim moved his machine gun businesses into European factories,[1] he had no need of the export licences that are necessary today.

Once the manufacture of military equipment became a large industrial and technological activity, and particularly after the First World War, the sale of such equipment by the private sector to foreign governments became a politically contentious activity. The profits made by munitions companies from the manufacture of the equipment associated with the slaughter of the First World War prompted a movement in the UK and elsewhere that the state should regulate defence exports and even that governments, not the private sector, should be responsible for making as well as using weapons. Article VIII of the Treaty of Versailles read:

> The members of the League [of Nations] agree that the manufacture by private enterprise of munitions and implements of war is open to grave objections. The Council shall advise how the evil effects attendant upon such manufacture can be prevented, due regard being had to the necessities of those Members of the League which are not able to manufacture the munitions and implements of war necessary for their safety.[2]

George Bernard Shaw's play *Major Barbara* and Philip Noel Baker's book *The Private Manufacture of Armaments* attracted significant attention, as did the trenchantly-titled 1934 American work of HC Engelbrecht and FC Hanighen, *Merchants of Death*.[3]

Yet, in the inter-war period, efforts to control defence exports stumbled in the face of international disagreements about their possible forms. Moves to reduce the role of the private sector in defence also had little success, but not because the state sector was thought less effective and efficient. As Lewinsohn explained in a work published in Paris in 1936:

> The argument put forward by the champions of free enterprise runs on these lines: Supposing the armaments works are nationalised today, the production of raw materials will be taken over tomorrow, for they are necessary to the manufacture of arms. The day after tomorrow it will be the turns of the textile industry and the victualling trade, for they too are essential parts of the war economy. It will never be possible to draw a line. The nationalisation of the arms business would be only the first step in the nationalisation of all industry.

He went on to show that the same kind of argument was used against export controls on defence items. As an aside, the UK was a significant beneficiary of the limited control over exports and the challenges of dealing with what today is defined as dual-use technology: the development and fielding of the Chain Home radar system that did so much to save the UK during the Battle of Britain required the covert import of radio components from German-controlled Austria.

Thus, it was just the imminence of war and to stop Germany being able to buy British military goods, that the UK rushed through the Import, Export and Customs Powers (Defence) Act Munitions Control Act of 1939,[4] legislation that remained in place without comprehensive review until the 1990 Import and Export Control Act and then the Export Control Act of 2002. After the Second World War, the British government continued to control the export of defence goods through licensing and customs arrangements, but was not particularly involved in supporting the export of British military goods except as a security-related activity to close allies. Then, from 1966, things changed.

The background was that the United States, which had supplied large amounts of military equipment to NATO European states as military aid, decided in the early 1960s that those countries had recovered sufficiently from the Second World War that they could afford to buy their own kit. Thus, the US government moved from being an aid donor to an active supporter of defence exports to its allies. Significantly, the UK had concerns about losing market share, not least in the military aircraft market.

In addition, many types of military equipment, not least in the aerospace sector, were becoming more and more expensive to develop. Sales beyond national markets appeared valuable both as a means to spread fixed development costs over larger production runs and also as a means to keep defence factories busy when national needs had already been met. For the UK, struggling to balance its trade in goods and services, defence exports could also be a valued source of foreign exchange.

These were the considerations that led the British government, on the basis of a recommendation of an enquiry led by a senior industrialist, to move in 1966 to being an active promoter of defence exports, as well as the regulator of such transactions, through the establishment of the Defence Sales Organisation, later (in 1985) renamed the Defence Export Sales Organisation (DESO) and from 2007 the Defence and Security Organisation (DSO).

DESO and the defence & security organisation

The original Defence Sales Organisation/DESO survived multiple changes of government in the UK with its duties of researching national arms markets and their decision-making systems and processes, identifying sales opportunities, helping companies with their bids for overseas work, representing the UK in arms exhibitions, arranging ministerial visits and meetings, and other governmental and military activities in support of major sales campaigns. London-based, but outside the Ministry of Defence's main building in Whitehall, it also posted small numbers of staff in important and potentially-important markets.

Its staff comprised mainly civil servants supplemented by some military staff to provide advice and specialist expertise, as well as being interlocutors for potential customers who wanted to speak primarily to uniformed personnel like themselves. A particular feature from the beginning was that its head should be a business person with significant experience in government-related exporting, and thus it was necessary for industry to fund a significant element of the salary that such a person would demand.

Organisationally DESO was part of the MoD. The integration of export sales efforts with UK export control concerns was through a small office in MoD Whitehall.

DESO's first 40 years were marked by periodic controversy and variation in the enthusiasm of defence ministers to play a personal role in the support of defence exports. But, as noted, the organisation survived in basically its original form and roles until 2008. At that point, Gordon Brown, who had become prime minister and who had a reputation for not being a defence industry enthusiast (unless it was building ships in his constituency),[5] ordered that DESO should be reduced in size and moved from the MoD to the Department of Trade and Industry (where it was renamed Defence and Security Organisation (DSO)).

This caused concern, not least in the defence industrial sector, in part because of concern that senior people from potential customers would want to be meeting defence rather than trade and industry ministers, and the former might be harder to get on board if they did not have direct responsibility. In the event, this proved to be not too much of an issue: military secondments to DSO continued and, from 2010, Prime Minister David Cameron was an individual keen that his ministers should do all possible to support defence exports. A DSO move back into MoD was considered but not held necessary, especially as 'Security' exports rather than defence items were gaining a larger place in UK export figures.

The justification for defence exports was and is threefold:

> Most obviously, defence exports, like all exports, generate employment, prosperity and tax revenues in the UK.
>
> Second exports, like collaborative defence projects with peer or near peer partners, supported the continuing survival of defence industrial capabilities in the UK in an era when national orders alone were often not enough to provide continuous work for defence industrial assets. For some systems and sectors, a capable national defence industry in turn increased the UK's capacity to use its forces as it saw fit without having to rely too much on external sources of supply. National kit could also be more easily modified for short notice operations in unforeseen theatres.[6]
>
> Third, exports provided the basis for long-term security relationships with other governments, increasing the British potential for international influence. Exports did not assure influence, but they did provide the opportunity to communicate at senior military and even political level with customer governments. It is in this way that defence exports were linked to the defence engagement domain.

UK export activity and performance

British exports can be assessed against a variety of criteria, including success compared with peer states, support for wider foreign, security and defence policy, and ethical considerations.

An initial challenge in this area is that those collecting statistics on the value of defence exports vary in what they seek to measure and have access to different information that is, in one way or another, partial. Not surprisingly they come out with very different numbers. Some of the important variables are whether orders are measured, or the values of actual deliveries, and what should be included within a 'defence' export: under its Al Yamamah commitments in Saudi Arabia, British companies have arranged English language tuition and built houses at new bases for Saudi personnel. Should the value of these activities be included as defence items, or included because they were integral elements of a capability package centred on the delivery of combat aircraft?

Table 6.1 below gives figures from the US Department of State in its long-standing web-based publication, World Military Expenditures and Arms Transfers.[7] It reports the arms transfers of 170 countries in billions of constant 2014 dollars. Each country's figure varies from year to year but over a decade the UK slightly leads France and Germany. One the other hand, UK sales are shown as having fallen since the 2004–2006 period. All three are clearly in the same league.

As a second source, the European Union assembles data from member states on the number and value of licences they have issued for items on the 22 sections of the Common EU Military List. Table 6.2 gives three years' figures which show the UK as having issued licences with a lower value than France and Germany. However, licences issued do not mean that an actual delivery has occurred. Countries may opt not to proceed with intended orders.

Table 6.1 Arms exports of goods and services of selected states (US Department of State)

	2004	2005	2006	2007	2008	2009	2010	2011	2012	2013	2014	Mean
UK	4.0	4.5	5.6	2.5	2.7	2.9	3.2	3.3	2.7	3.0	3.0	3.4
France	6.9	3.3	2.5	3.0	1.6	1.7	3.0	3.3	3.9	4.0	5.1	3.5
Germany	2.6	2.3	2.5	3.5	4.5	4.7	4.5	2.6	1.9	1.3	2.1	3.0

Source: World Military Expenditures and Arms Transfers (WMEAT), 2016.

Table 6.2 EU data on value of military list licences issued in 2014, in 2014 euro[8]

	2014 euro	2013 euro	2012 euro
UK	2,585,633,021	5,232,124,959	2,664,113,244
France	3,846,640,000	9,538,437,961	13,760,313,539
Germany	3,973,800,137	5,845,628,422	4,703,969,983

Source: European Union, Annual Report on Arms Exports, 2016.

The Stockholm International Peace Research Institute has its own method of recording the value of arms deals (using the concept of Trend Indicator Values (TIVs)). A table covering seven years is laid out in Table 6.3 below. This suggests that both France and Germany sold more than the UK over a seven-year period.

A fourth source of what have to be called estimates is the US Congressional Research Service, with its annual report on *Conventional Arms Transfers to Developing Nations*. Table 6.4 summarises agreements, i.e. orders rather than deliveries, and, like SIPRI, finds that the UK has been a significant seller, albeit not on the same scale as France and Germany.[9]

However, British figures from DSO and approved by the Office of National Statistics tell a very different story, and report significantly higher British defence export orders than those noted by the State Department, SIPRI, or the Congressional Research Service. DSO asserts that, in terms of orders the UK is normally the second highest defence exporter after the United States, although it recognised that in 2015 France jumped ahead for a year at least because of major contracts with Qatar and Egypt. However, it also recognised that it had only partial information about other states' activities, as Table 6.5 reveals.

While there is plenty of variation regarding export values, there is complete consensus that the UK's main market is the Middle East and Saudi Arabia in particular, with its fleets of Tornado, Hawk, and Typhoon aircraft. This dependence presents political and economic risks for UK defence industry in general and the combat aircraft in particular. However, given the volume of Saudi spending and the closed nature of the large Chinese and Russian defence markets, it is not an easy dependence from which to escape.

However, as the UK's national reports on military list export licences indicate, the UK also has a large number of countries that are its small, medium, and large defence customers. Of significance for the defence engagement section to

Table 6.3 SIPRI data on value of UK defence exports 2010–2016

	SIPRI trend indicator values 2010–2016
UK	8,761
France	11,226
Germany	11,993

Source: Stockholm International Peace Research Institute.

Table 6.4 Worldwide arms transfer agreements 2008–2015, in millions of US$ 2015

	2008–11	*2012–15*
UK	4,000	10,800
France	23,200	15,900
Germany	9,600	15,900

Source: Theohary, Conventional Arms Transfers to Developing Nations, 2008–2015, 2016.

Table 6.5 UK defence exports (based on orders) 2006–15, in current £billion

Year	£bn
2006	5.5
2007	9.7
2008	4.3
2009	7.3
2010	5.8
2011	5.4
2012	8.8
2013	9.8
2014	8.5
2015	7.7

Source: Stockholm International Peace Research Institute.

follow in this chapter, the UK provided what SIPRI classifies as 'Major Conventional Weapons' (but which are actually either defence platforms or major subsystems of such systems) in the 2010–2016 period to more than 30 countries. Because of the support needs of most of this equipment, these deliveries require continuous contact between the users and relevant UK authorities.

The aerospace dominance of UK exports is unsurprising given that the UK's submarine industry is entirely nuclear in focus, with no export opportunities. On land the government has allowed the land vehicles industry to shrink: without programmes for UK land system development, the UK's design and production capacity fell. The last stages in this were arguably the UK withdrawal from the Boxer programme in 2003, the failure to select BAE Systems for any major land programme this century, and the closure of the tank plants in Leeds and Newcastle. Ship sales have been rare, involving either smaller vessels or retired and refurbished RN Type 23 frigates. A bright spot has been the export success of MBDA, whose sales of complex weapons matched domestic sales by 2016.

Export controls policies

Although there are multiple benefits for the exporter from defence sales, the UK does not sell anything to anybody, but has a formal export control policy, and processes and organisation for its application.[10]

A fundamental point is that the UK's export controls are comprehensive and effective in that they cover most relevant activity.

Lists

There are extensive Military and Dual-Use national lists of goods whose content has been agreed at the European level as requiring an export licence. Most items on these technically national lists also reflect the content of the Wassenaar Arrangement. It is an exporter company's responsibility to be aware if exported products are covered and subject to licences.

Activities needing licences

Listed goods leaving the UK obviously need licences, even if an item is remaining in UK corporate ownership and control because it is being sent to an exhibition. The export of intellectual property also requires licences. Clearly controlling the export of information is a huge challenge in the digital age and when so much can be stored on an innocent-looking pen drive. In addition, UK-based personnel and firms (brokers) arranging the movement of listed goods from a foreign state to a third country must request and be granted an export licence (this practice is most common with regard to light arms). Further, the Export Control Act of 2002 also required British citizens based overseas who are engaged in the defence trade to get British approval for their international deals.

Like that in other Western states, UK legislation includes a 'catch-all' clause which says that, if an exporter suspects that any export sale is contributing to a military purpose, it should seek a licence. The particular concerns of this clause in practice relate to when the sale is to a country under an arms embargo or when the military use could relate to a 'non-conventional' weapon of mass effect (biological, chemical, or nuclear) or to a ballistic missile.

The one area where effective legal regulation has so far proved impractical concerns the 'export' of personnel by private security companies as part of the delivery of contracts to overseas customers. This difficult topic is beyond the scope of this chapter, but such people rarely if ever travel with their weapons.

Processes and organisation

The UK has an all-electronic set of processes for the submission of applications for the various sorts of licences that can be issued. Firms submit forms to the Export Controls Unit of about 100 people in the Department for International Trade (its name from 2016). Larger and even medium-sized firms have to employ specialised staff familiar with the information and processes required to deal with this unit. The ECU staff coordinate and implement consultation with other government departments and interests, including the Foreign and Commonwealth Office, the MOD and its agencies including DSTL, and the intelligence services, and then issue a decision. They are under pressure from companies to do this quickly and the unit reports average response times, which are normally measured in days rather than weeks or months.

There is also a special process relating to the release of classified ('secret' in plain English) rather than export-controlled information. In order to win a military sale, it is often necessary that the prospective customer be provided with information about the system's performance and other attributes. To reduce the chance of UK forces being put at risk by the release of this information to a third party, the company must apply for permission from the MoD through completion and submission of a Form 680.

All UK companies must obtain MOD Form 680 approval in order to release information or equipment classified OFFICIAL-SENSITIVE and above to foreign entities.[11]

The Form 680 and the export licence processes are clearly related but technically separate. The MoD may approve the release of information to a prospective customer only for the government to later decide that it should not allow the actual export of the product.

Export controls also require customs agencies of integrity and capability. Although there can be no certainty that smuggling is not occurring, the British customs service is believed capable and the deliberate unauthorised export of defence equipment and technology is not recognised as a significant problem. Espionage and the theft of intellectual property, of course, remain challenges.

Policy

Clearly a licence cannot be granted if the UK has committed to an 'arms embargo' on a specific state. However, this may not be an entirely straightforward matter since the UK rarely applies embargoes on a national basis. They are normally applied under an EU, NATO, or United Nations framework and the specific terms of any embargo may vary from case to case. The government publishes a full list of embargo arrangements and other structured restrictions.[12]

British export controls reflect substantial agreement on principles among EU members on these matters since the mid-1990s. EU states agreed in 1998 that they would take into account eight factors when considering arms exports to a specific destination and a formal 'Common Position' on military technology and equipment transfers was agreed in 2008.[13] These eight elements are specified below. Predictably they are similar to, but not identical to, the concerns noted in the Arms Trade Treaty of 2016 to which the UK is also a party. In particular, the latter has more stress on organised crime, which was not as large a concern in 1998 as it became 15 or so years later.

Under the Common Position, governments refusing an export report their stance to their EU partners. The latter, if they subsequently agree to a similar deal with the original customer, must report their choice to the others. The system thus discourages states from approving proposals on the grounds of believing that, if one says no, another state will step in. When a deal is refused, reference is made to the specific factors in play. The eight considerations are:

1   Respect for the international obligations and commitments of Member States, in particular the sanctions adopted by the UN Security Council or the European Union, agreements on non-proliferation and other subjects, as well as other international obligations;

2   Respect for human rights in the country of final destination, as well as respect by that country of international humanitarian law;

3   The internal situation in the country of final destination – Member States will not allow exports that would provoke or prolong armed conflicts or aggravate existing tensions or conflicts in the country of final destination;
4   The preservation of regional peace, security, and stability;
5   The security of Member States and of territories whose external relations are the responsibility of a Member State, as well as that of friendly and allied countries;
6   The behaviour of the buyer country with regard to the international community, as regards in particular its attitude to terrorism, the nature of its alliances, and respect for international law;
7   The existence of risk that the military technology or equipment will be diverted within the buyer country or re-exported under undesirable conditions;
8   The compatibility of the exports with the technical and economic capacity of the recipient country, taking into account the desirability that states should meet their legitimate security and defence needs with minimal diversion of human and economic resources for armaments.

The interpretation and application of these commitments, and the national legislation that legalises the licensing system, remain a matter for national judgement and action, so the UK still enjoys significant discretion and a requirement for the exercise of judgement. The government maintains that it is cautious and rigorous.

> A government spokeswoman said: "The government takes its arms export responsibilities very seriously and operates one of the most robust arms-export-control regimes in the world.
> "We rigorously examine every application on a case-by-case basis against the consolidated EU and national arms export licensing criteria and are satisfied that extant licences are compliant with these."[14]

Its critics, especially those in the Campaign Against the Arms Trade (CAAT), maintain a very different stance, and in 2017 (unsuccessfully) pursued a judicial review to halt UK defence sales to Saudi Arabia. Clearly there are ethical and emotional issues around defence exports where consensus is likely to be elusive in an open society such as that of the UK. On the one hand, there is the UN Charter assertion of all rights to self-defence: given the limited possibility for most UN states to produce much of their needed defence equipment, this implies that all states have a right to import arms they feel they need. There is also the UK strategic interest in employment, defence industrial capability, and political influence. On the other hand, many countries have governments that do not treat their citizens well, as the World Bank's Governance Indicators underline[15] and scholars such as Acemoglu and Robinson demonstrate.[16]

Government support for defence exports, like any policy stance, clearly brings risks that can have an extended life. The British government faced embarrassment and difficult choices about whether to halt arms deliveries to Saudi Arabia in 2016 when there was strong evidence that Saudi Arabia had used UK-origin cluster munitions in its campaign against Yemeni rebels. These weapons had actually been delivered at latest in 1989, before the UK-ratified Treaty forsaking their use.

Another hazard is that the government can be at least indirectly associated with corrupt behaviour. Many governments are, unfortunately and tragically for the societies they control, much involved with corruption. The defence trade, with its high-value deals, is always a tempting target and many companies and purchasing governments have over decades been caught up in corruption scandals. By 2017 the British Bribery Act represented a significant deterrent and the major defence companies had rigid ethics codes to guide and control their staff. BAE Systems developed and introduced its transparent systems with the help of a distinguished judge, Lord Wolfe, after many payment details of its 30-year relationship with the Saudi government had emerged.[17] Similarly, Lord Leveson accepted in 2017 that there had been 'cultural change' at Rolls Royce since 2010 when the board had allegedly been informed of improper payments.[18]

Corruption in government life, indeed in the private sector as well, is not a problem with a solution: it is an ongoing challenge that needs continuous effort to minimise its incidence and impact. Active British government involvement with defence sales promotions should work to make it less of a feature. The UK government normally deals directly with the customer government, whereas much corruption is associated with companies' use of local agents to win favour for their offering with local officials.

Exports in the shaping of UK requirements

Until the 2010 Strategic Defence and Security Review, British practice on requirements was very much that UK forces specified the equipment capabilities that they judged necessary. If any other governments subsequently wanted to buy the equipment that emerged to meet those requirements, it would be an incidental benefit. Even once the government began to recognise the value of considering exportability in the shaping of UK requirements and their solutions, in practice very little notice was taken of this.

Then the financial crisis of 2009 hit the UK economy and the defence budget. The MoD, like other ministries, was directed to contribute to the 'prosperity agenda' and began to take seriously the specification of requirements for solutions in a manner which increased the possibility for exports. While the idea of reducing the capability ambition of British equipment was unappealing, there might be possibilities for a modular approach enabling a less complex version of a system meeting the needs of foreign customers. A related possibility was of more collaborative arrangements with peer states with similar needs to the UK. This thinking was first applied with regard to the replacement for the Navy's Type 23 frigate, eventually designated the Type 26. However, external interest initially proved elusive, although Australia did select BAE Systems in 2016 as a competitor for its Global Combat Ship programme with a modified Type 26 to be built in Australia.[19] The Navy's demanding specification for the ship had increased the expected cost to £1 billion per vessel.[20] Thus by 2017 it seemed settled that only eight Type 26s could be afforded, but that an additional number of less sophisticated vessels, the 'Type 31's, would be built for naval missions where there was a less demanding threat environment. Export potential for the latter type would appear greater.

Despite the policy commitment to include export prospects in the specification of requirements, the readiness of UK forces actually to define their needs, taking account of what other states might also want, remained an open question.

Defence exports and training

Another element of UK defence preparations where exports intruded was training. Training is a multi-layered activity: individuals often have to learn separately how to operate their equipment (fly the aircraft or drive the tank), then they have to be taught how to fight with the equipment on an individual and, where relevant, crew basis. In addition, collective training is often needed where equipment is used in conjunction with other units, such as a brigade level exercise.

By 2017, responsibility for much training activity in the UK at the individual level had been privatised: companies had been contracted to provide the needed training through a long-term contract that justified the investment they needed to make in equipment and employees. The Military Flying Training System contract was a case in point.

An issue was whether the MoD would be ready to size the provision of training, and whether the Treasury would allow the MoD to size the provision of training, so as to enable the training of foreign personnel associated with (uncertain) export sales. In the Treasury's eyes, the MoD had to make a business case, and be ready to accept the commercial and financial risk involved, for the extra spending commitments involved.

Clearly all foreign customers for defence equipment would need training, but in many cases part of the appeal of a British system lay in its potential for foreign countries to gain access to UK ways of doing things through training alongside UK personnel. One senior company representative told us of one successful naval sale that was seen by the customer as enabling its navy to stay in regular contact with the Royal Navy.

This is not an issue with a black or white response. The government recognises that, if it cannot offer UK government-endorsed training, sales efforts will be damaged. On the other hand, it can maintain only limited spare capacity for overseas students that may not materialise.

From defence exports to defence engagement

UK training, however, is not just a means to support defence exports. It also supports the purposes associated with the policy line defined in 2015 as Defence Engagement, but having other designations before that.

> Defence engagement is the use of our people and assets to prevent conflict, build stability and gain influence.... Defence engagement projects influence, promotes our prosperity and helps to protect our people.[21]

Going back to the 1998 SDR and its commitment to the UK with force projection and being a force for peace and stability in the wider world, and even before that as being concerned with the promotion of UK influence, not least in countries that were formerly part of the British Empire, the British defence establishment has long undertaken a range of tasks, of which a sample are listed below.

1   British military training and education institutions have long taken a significant number of students from overseas, intending that they should leave with familiarity and appreciation both for UK military expertise and the UK way of life. The Army's officer cadet school at Sandhurst, the Advanced Command and Staff course, and the course at the Royal College of Defence Studies continue their tradition of taking a large number of overseas staff, with the latter particularly seeking to bond individual and overseas participants;
2   Small but increasing numbers of UK military and contracted personnel have been deployed in packages to provide specialised military training to overseas forces;
3   Education in security sector governance and management has been organised in foreign countries of interest to the UK and where there was felt to be a good chance of a favourable reception for such work;
·4   British military attaches and related personnel have been sent as advisers inside the defence establishments of foreign states with major change programmes;
5   Naval vessels visit ports in friendly overseas states, prompting high-level contact with the local government;
6   Ad hoc exercises have been conducted with friendly states.

In the years following the SDR of 1998, much of this was organised under the heading of Defence Diplomacy, and in many ways 'International Engagement' is Defence Diplomacy with a lot of extra elements and given a new badge.

The formal specification of Defence Engagement as a funded, core task of the MoD was a feature of the 2015 SDSR, and in 2017 the government published an International Defence Engagement (IDE) Strategy document which spelled out first that IDE was to support all three objectives of UK national security: the protection of the British people, the promotion of prosperity, and the projection of UK global influence.

The Strategy document made clear that IDE was a large toolbox with many tools and multiple uses, including everything from a readiness to work with close allies, including on collaborative development projects, to providing training and support to governments emerging from civil conflicts in southern Africa.

> Through capacity building activity, including security sector reform, short-term training teams, international training and exercises we will focus on building partner capability.[22]

The earlier stress on overseas students on UK-based courses remained:

> UK training and education is in high demand. All three services and Joint Forces Command provide training in the UK to improve individual and collective skills, in anything from platoon tactics to cyber defence. We also provide high prestige education courses in the UK, such as places on Initial Officer Training at the Royal Naval College, Dartmouth, the Royal Military Academy, Sandhurst, and RAF College, Cranwell, and deliver world-leading maritime training through the Royal Navy's FOST (Flag Officer Sea-Training) organisation. Places offered to international students at the Royal College of Defence Studies and other Defence Academy courses are highly prized. Alumni from these countries frequently rise to senior government or military posts, providing the UK with enduring influence and cementing our bilateral relations with partners. Both training and education also increase understanding of UK and international norms on priority areas such as human rights, transparency and the rules-based international order, while also helping to develop enabling capabilities and institutions. We are delivering a greater range and volume of International Defence Training, and have increased the number of international places we offer on our highest profile courses by more than 10% in 2015–16 and 2016–17. We have piloted a new International Senior Strategic Leadership Programme and International Strategic Planning courses, introduced a Royal Navy International Officers' Course, and increased the number or ratio of international student places on a wide range of other courses.[23]

Many defence engagement activities with the development world are designed both to strengthen the governance and effectiveness of local military and security forces (thus making it less likely that external intervention in conflict will be needed) and to facilitate UK capacity to intervene from the UK in cases where conflict does occur and intervening British forces need host nation support of one sort or another.

As implied, there are three different if related realms of defence engagement: the first and often the most straightforward is to improve the professional operating skills of foreign forces, a task which needs serving or retired military personnel. The second is to strengthen the integrity and governance of foreign militaries so that corruption is reduced and subjugation of the military to political authorities is enhanced. The third is to improve the management of defence resources to minimise the waste of scarce defence funds. Responsibility for supporting in the latter two areas often falls to civilians with a good blend of government experience and intellectual knowledge.

To support International Defence Engagement, the MoD has put in place some extra resources, not least in the area of defence attaches, and a suite of conceptual and organisational arrangements.[24] It could be argued that the reorganisation of the Army into basically two divisions has been much affected. While 3 Division is prepared as the division with readiness for large-scale or high intensity operations, 1 Division is providing the capability and expertise for the training and other activities associated with defence engagements. The British Army standing presence in Africa includes advisory teams in South Africa and Sierra Leone, support to the Mine Action Training Centre in Nairobi, and the British Peace Support Team, also in Nairobi, which leads on UK activities in east Africa to promote security sector reform and enhance peacekeeping capabilities.[25]

Such a stress on defence engagement for the Army is not without its downside. It may reflect an excessive concern to demonstrate that UK armed forces are 'busy' rather than being 'ready' for a major operation. There is also concern that those deployed on Defence Engagement tasks will feel part of the second class rather than the elite element of the army. Happily, such issues loom less large for the Navy and the RAF. The former in particular has a long-standing tradition of foreign port visits (including cocktail parties on the helicopter landing area).

The IDE Strategy asserts the mutually-supportive nature of defence exports and defence engagement: defence diplomacy and other contacts with international militaries can facilitate UK exports, while defence exports enhance military-military dialogues. Thus, a significant paragraph in the strategy document addresses the macro-economic benefits of defence exports to the UK and their place in International Defence Engagement.

> Defence exports make a significant contribution to the UK's national prosperity (worth £7.7Bn in 2015) and reinforce our key strategic international defence relationships through increased military-military dialogue and interoperability. UK defence exports also help secure the UK industrial base and maintain sovereign capabilities. The SDSR made support to exports an integral part of MOD's core business. Defence engagement makes an important contribution to the understanding of economic opportunities in countries of interest and how to take them forward. It supports government export campaigns, including through the work of its Defence Attachés or attendance at exhibitions, and on occasion forms part of the package offered, including training or maintenance support.[26]

In terms of diplomatic access, the export of a major defence platform is a reason for intergovernmental discussions, sometimes at a senior level, between

UK and customer officials, over decades. It gives an opportunity for communication and potentially influence on a range of subjects well beyond the defence domain.

While the International Defence Engagement brand was launched in the 2015 SDSR, emphasis on it increased as the vote to leave the European Union pushed the government to talk with extreme confidence on the future of 'Global Britain', where Defence Engagement was to have a prominent place.

> As we leave the European Union we will be more prominent on the world stage than ever: an outward-facing global partner at the heart of international efforts to secure peace and prosperity for all our people. The UK will continue to be one of the principal security and defence actors in the world, demonstrated by our ongoing commitment to spend 2% of GDP and 0.7% of GDP on international aid.[27]

While the document does not fully reconcile the above with another statement in its text ('the UK is unlikely to fight overseas against a sophisticated adversary on its own'), the Brexit choice will enhance pressure on the MoD to sustain its defence engagement efforts.

Conclusion

Defence exports have been both a notable element in UK prosperity and an important factor in the country's capacity to sustain a wide-ranging set of defence industrial capabilities, which in turn are a key element in the credibility of the UK to be a major and independent military power. The ability to provide useful defence equipment to other states also offers possibilities for access to and so influence with the highest levels of customer governments, while a state dependent on external suppliers for its defence equipment has significant constraints on its freedom of action.

Looking forward, the worry must be about the availability of UK products for sale. The core figures about reduced research and development spending even since 2001 have clear implications for the range of defence products that can be in the British defence shop window in a decade's time. In real terms, development expenditure by the MoD fell by 56 per cent between 2001/2002 and 2014/2015,[28] a period when US State Department figures show a significant and steadily rising total of British defence imports as per Table 6.6.

Table 6.6 US State Department figures on UK defence imports in constant 2014 US$ billion[29]

2004	2005	2006	2007	2008	2009	2010	2011	2012	2013	2014	*Average*
9.6	8.5	9.1	9.3	11.0	11.5	11.6	12.1	12.1	11.6	9.9	10.6

Source: World Military Expenditures and Arms Transfers (WMEAT), 2016.

The more that the UK relies on imported systems, the less it will be able to offer arms supplies as a route to international influence. It will thus have to rely more on the professional expertise and reputation of its people as a means of impressing other governments and armed forces.

Defence exports and defence engagement should and normally do have a mutually-reinforcing relationship. Ministerial visits and exchanges most obviously serve both, providing the top-level link to support a range of activities including export drives and training offers. Clearly, stresses are almost inevitable when a customer uses equipment in a non-conventional war, especially if that use does not have the character of clear self-defence against an external aggressor. In 2016 it was Saudi behaviour in Yemen that was a problem. Almost 20 years earlier it had been Indonesian military operations in East Timor, with the suspicion that Hawk jets had been used for internal security purposes.

An inevitable feature of defence engagement as a set of diverse activities is that, in many cases, it is difficult to measure success and to specify precise outcomes that can be attributed to engagement activities. They make it more likely that the UK will enjoy influence in the wider world, if only because British authorities are in regular contact with the militaries and even political figures of a foreign state. They also make it more likely that the governance of security sectors of targeted states will improve, but there are no assurances of these results or simple ways to measure them. It can be expected that two similar programmes of activities would have different consequences in different countries and contexts.

Finally, the 2015 SDSR spoke of UK defence as being 'international by design', a vague phrase that was meant to cover a wide range of activities, including those in the acquisition and operational spaces. However, the UK's approach to defence engagement and exports is another specific aspect of the 'international by design' label.[30]

Notes

1 CJ Chivers, *The Gun* (London, Penguin, 2010), Chapter 3; Anthony Smith, *Machine Gun: The Story of the Men and the Weapon That Changed the Face of War* (London, St. Martins Press, 2004).
2 Library of Congress, www.loc.gov.
3 HC Engelbrecht and FC Hanighen, *Merchants of Death: A Study of the International Armament Industry* (New York: Dodd, Mead, 1934).
4 See Tamotsu Aoi, 'Historical development of export control systems in selected countries and regions', CISTEC (Center for Information on Security Trade Control), April 2016 www.cistec.or.jp/english/service/report/1605historical_background_export _control_development.pdf, accessed 19 July 2017.
5 BBC News, 'Gordon Brown Urges Carrier Work at Rosyth', *BBC News*, 2 November 2010, www.bbc.co.uk/news/uk-politics-11662051, accessed 19 July 2017.
6 See Trevor Taylor, John Louth and Henrik Heidenkamp, 'Industry and the Military Instrument' in Adrian Johnson (ed.), *Wars in Peace* (London, RUSI, 2015).
7 US Department of State, 'World Military Expenditures and Arms Transfers 2016', www.state.gov/t/avc/rls/rpt/wmeat/2016/index.htm, accessed 15 February 2017.
8 European External Action Service, www.eeas.europa.eu/non-proliferation-and-disarmament/arms-export-control/index_en.htm, accessed 19 December 2017.

9 See Catherine A Theohary, 'Conventional Arms Transfers to Developing Nations, 2008–2015', *Congressional Research Service*, 19 December 2016, https://fas.org/sgp/crs/weapons/R44716.pdf, accessed 24 January 2017.

10 Jon Lunn, 'The UK Legal and Regulatory Framework for UK Arms Exports', London, House of Commons Library, 4 September 2017, http://researchbriefings.parliament.uk/ResearchBriefing/Summary/SN02729#fullreport, accessed 4 October 2017; see also the Guidance from the Department for International Trade at Gov.uk, www.gov.uk/guidance/export-military-or-dual-use-goods-services-or-technology-special-rules, accessed 4 October 2017.

11 Ministry of Defence, 'Ministry of Defence (MOD) Form 680 applications', www.gov.uk/guidance/mod-f680-applications, accessed 1 January 2018.

12 See Department for International Trade, 'Current arms embargoes and other restrictions' www.gov.uk/guidance/current-arms-embargoes-and-other-restrictions#history accessed 01 January 2018.

13 European External Action Service, www.eeas.europa.eu/non-proliferation-and-disarmament/arms-export-control/index_en.htm; and Council Common Position 2008/944/CFSP, *Official Journal of the European Union*, 8 December 2008 http://eur-lex.europa.eu/LexUriServ/LexUriServ.do?uri=OJ:L:2008:335:0099:0103:EN:PDF, accessed 26 July 2015.

14 Jamie Doward, 'UK weapons sales to oppressive regimes top £3bn a year', *Guardian*, 28 May 2016, www.theguardian.com/world/2016/may/28/uk-weapons-sold-countries-poor-human-rights-saudi-arabia, accessed 17 July 2016.

15 The reader can use the interactive system at World Bank, Worldwide Governance Indicators, http://info.worldbank.org/governance/wgi/#home, accessed 19 July 2017.

16 Daron Acemoglu and James A Robinson, *Why Nations Fail: the origins of power, prosperity and poverty* (London: Profile Books, 2012).

17 See BAE Systems, 'Our Response to the Wolfe *Committee Recommendations*, accessed through www.baesystems.com/en/download/1434568989614.pdf, 4 October 2017.

18 Graham Ruddick and Graeme Wearden, 'What did Rolls-Royce directors know about bribery scandal? "No comment"', *Guardian*, 19 January 2017, www.theguardian.com/business/2017/jan/19/what-did-rolls-royce-directors-know-about-bribery-scandal-no-comment, accessed 15 February 2017.

19 George Allison, 'BAE Systems sign Type 26 Frigate contract with Australia', *UK Defence Journal*, 1 September 2016, https://ukdefencejournal.org.uk/bae-systems-sign-type-26-frigate-contract-australia/, accessed 9 December 2016.

20 In July 2017, the Ministry of Defence contracted for the first three T.26 ships. The contract was valued at £3.7 billion: Royal Navy, 'Multi-billion pound deal signed for first three new Type 26 frigates', 2 July 2017, www.royalnavy.mod.uk/news-and-latest-activity/news/2017/july/02/170702-deal-for-t26-frigates, accessed 7 September 2017.

21 MoD, 'International Defence Engagement Strategy' (London: TSO, 2017), p. 1.

22 Ibid., p. 12.

23 Ibid., p. 13.

24 Ministry of Defence, *The United Kingdom's Defence Engagement Strategy, 2017* (London: MoD, 2017).

25 www.army.mod.uk/operations-deployments/22724.aspx.

26 MoD, 'International Defence Engagement Strategy' (2017), p. 16.

27 Ibid., p. 3.

28 See Trevor Taylor, 'Supplementary written evidence', on Defence Acquisition and Procurement submitted to the Defence Committee, 17 January 2017.

29 US State Department, 'World Military Expenditures and Arms Transfers 2016', Table II, IV, www.state.gov/t/avc/rls/rpt/wmeat/2016/index.htm.

30 SDSR 2015, p. 1.

Bibliography

Acemoglu, D and Robinson, JA (2012), *Why Nations Fail: the origins of power, prosperity and poverty* (London: Profile Books).

Allison, G (2016), 'BAE Systems sign Type 26 Frigate contract with Australia', *UK Defence Journal*, 1 September 2016, https://ukdefencejournal.org.uk/bae-systems-sign-type-26-frigate-contract-australia/, accessed 9 December 2016.

Aoi, T (2016), 'Historical development of export control systems in selected countries and regions', CISTEC (Center for Information on Security Trade Control), April 2016 www.cistec.or.jp/english/service/report/1605historical_background_export_control_development.pdf, accessed 19 July 2017.

BAE Systems, 'Our Response to the Wolfe *Committee Recommendations*, www.bae systems.com/en/download/1434568989614.pdf, accessed 4 October 2017.

BBC News (2010), 'Gordon Brown Urges Carrier Work at Rosyth', *BBC News*, 2 November 2010, www.bbc.co.uk/news/uk-politics-11662051, accessed 19 July 2017.

Chivers, CJ (2010), *The Gun* (London, Penguin).

Council Common Position 2008/944/CFSP, *Official Journal of the European Union*, 8 December 2008, http://eur-lex.europa.eu/LexUriServ/LexUriServ.do?uri=OJ:L:2008:3 35:0099:0103:EN:PDF, accessed 26 July 2015.

Department for International Trade, 'Guidance', www.gov.uk/guidance/export-military-or-dual-use-goods-services-or-technology-special-rules, accessed 4 October 2017.

Department for International Trade, 'Current arms embargoes and other restrictions' www.gov.uk/guidance/current-arms-embargoes-and-other-restrictions#history, accessed 1 January 2018.

Doward, J (2016), 'UK weapons sales to oppressive regimes top £3bn a year', *Guardian*, 28 May 2016, www.theguardian.com/world/2016/may/28/uk-weapons-sold-countries-poor-human-rights-saudi-arabia, accessed 17 July 2016.

Engelbrecht, HC and Hanighen, FC (1934), *Merchants of Death: A Study of the International Armament Industry* (New York: Dodd, Mead).

European External Action Service, www.eeas.europa.eu/non-proliferation-and-disarmament/arms-export-control/index_en.htm, accessed 19 December 2017.

Library of Congress, www.loc.gov, for Treaty of Versailles.

Lunn, J (2017), 'The UK Legal and Regulatory Framework for UK Arms Exports', London, House of Commons Library, 4 September 2017, http://researchbriefings. parliament.uk/ResearchBriefing/Summary/SN02729#fullreport, accessed 4 October 2017.

Ministry of Defence (2017), *The United Kingdom's Defence Engagement Strategy, 2017* (London: MoD).

Ministry of Defence (2017), 'UK's International Defence Engagement Strategy', 17 February 2017, www.gov.uk/government/publications/international-defence-engagement-strategy-2017, accessed 21 February 2017.

Ministry of Defence, 'Ministry of Defence (MOD) Form 680 applications', www.gov.uk/guidance/mod-f680-applications, accessed 1 January 2018.

Royal Navy (2017), 'Multi-billion pound deal signed for first three new Type 26 frigates', 2 July 2017, www.royalnavy.mod.uk/news-and-latest-activity/news/2017/july/02/170 702-deal-for-t26-frigates, accessed 7 September 2017.

Ruddick, G and Wearden, G (2017), 'What did Rolls-Royce directors know about bribery scandal? "No comment"', *Guardian*, 19 January 2017, www.theguardian.com/business/2017/jan/19/what-did-rolls-royce-directors-know-about-bribery-scandal-no-comment, accessed 15 February 2017.

Smith, A (2004), *Machine Gun: The Story of the Men and the Weapon That Changed the Face of War* (London, St. Martins Press).

Theohary, CA (2016), 'Conventional Arms Transfers to Developing Nations, 2008–2015', *Congressional Research Service*, 19 December 2016, https://fas.org/sgp/crs/weapons/R44716.pdf, accessed 24 January 2017.

Taylor, T (2017), 'Supplementary written evidence', on Defence Acquisition and Procurement submitted to the Defence Committee, 17 January 2017: http://data.parliament.uk/writtenevidence/committeeevidence.svc/evidencedocument/defence-committee/defence-acquisition-and-procurement/written/44710.html, accessed 22 September 2017.

Taylor, T, Louth, J and Heidenkamp, H (2015), 'Industry and the Military Instrument' in Adrian Johnson (ed.), *Wars in Peace* (London, RUSI).

US Department of State (2016), 'World Military Expenditures and Arms Transfers 2016', www.state.gov/t/avc/rls/rpt/wmeat/2016/index.htm, accessed 15 February 2017.

World Bank, Worldwide Governance Indicators, http://info.worldbank.org/governance/wgi/#home, accessed 19 July 2017.

www.army.mod.uk/operations-deployments/22724.aspx.

7 Defence as skills and competencies

Introduction

As we have seen in earlier chapters of this book, the relationships that exists between government, its military, the British people, and the commercial businesses that operate in defence is complicated and multi-faceted. There is a political imperative to defence, commercial realities to be addressed, broader societal permissions for activities that occur under the banners of defence and security, as well as national, regional, and local economic considerations.

One of the constant discourses relating to defence in the UK is to consider the importance of technical and functional skills and competencies on myriad workforces. Analysts may describe the employees within defence industries as being significant as capability partners in a theatre of operation on the frontline, as for instance in Afghanistan. Others might prefer the notion of the broader defence industrial base as a critical national knowledge repository and an equipment and support hub able to surge outputs at moments of national crises. A different thinker might take the simpler perspective that defence businesses merely design, manufacture, and service equipment used by the armed forces, before being involved in its safe disposal, upgrade, or replacement. Yet common to each of these interlocking perspectives is the sense that the skills and competencies of the defence industrial worker are as important to defence as the skills and competencies of the soldier, sailor, airwoman or man; that the critical skills and competencies of defence reside with people who wear overalls on a factory floor as well as with those who dress in military uniforms.

Chapter objectives

By the end of this chapter the reader will understand:

1   The definition of skills and competencies within the context of UK defence;
2   The evolution of competency thinking in relation to defence reform;
3   The roles and purposes of commercial competencies-based thinking when applied to the defence sector;

4   The complexity of skills and competency thinking in relation to the UK defence system;

5   The challenge of the need for agility to a rigid, competence system.

Chapter structure

This chapter starts with a discussion on the definitions, roles, and significance of skills and competencies within defence in the UK. It goes on to articulate and debate the evolution of competency thinking and its relationship to defence reform, particularly from the Blair years onwards and the Smart Procurement Initiative of 1998. This leads to a discussion on the importance of people, as well as their labelled competencies, to the delivery of military effects and defence solutions. It concludes with a discussion on skills and competency thinking relating to the UK defence system, drawing-out the system's complexities and contingent understandings.

An interesting element of the skills and competencies narrative across defence is the perception that the skills in the defence industrial base, as well as military skills in the armed forces, somehow, self-regulate. In industry, when one company within the defence system makes redundancies, perhaps due to a perceived over-capacity or the lack of future military requirement, another defence business will absorb these skills and indeed many of the employees themselves. As a consequence, policy makers believe that defence industrial competencies stay alive and remain available to the nation-state, essentially frozen into some form of perpetual equilibrium underpinning our security. It is an act of governmental faith that:[1]

> A greater proportion of our overall defence and security business is available to industry than in any other major defence nation [delivered through] a highly skilled and flexible labour force.

How well-founded this faith is, is a topic we come now to discuss.

Skills and competencies within UK defence

Definitions

For the purposes of this book, a skill can be described as: 'An expertise, practiced ability, facility in an action or dexterity within a particular subject.' As can be imagined, with this as our definition there must be myriad skills across the UK defence enterprise and broader society within which it sits. Moreover, this notion of skills-base is hard to codify, measure, and manage. 'Competency', in contrast, can be defined as:

> The capacity to deliver a particular service for stated, pre-planned effect, through the application of skills or skills and know-how, plus the demonstrable ability to attest to this ability through the presence of a formal qualification or certification.

At first glance this might seem like the act of dancing on the head of a pin, but these distinctions are present and active across a highly managerialised defence sector. As we saw in Chapter 3, the managerialisation of defence is a constant since the end of the last century, with this applied significantly through notions of competency-based management. Indeed, one ex-government minister of defence attested as follows: 'I never heard of a defence competency until 1998. Thereafter, I heard of little else. I still don't know what they are, though.'[2] What this hints at for defence today we come on to discuss.

Defence competencies in an era of austerity

When thinking about defence competencies through the lens of the definition offered above, it is hard not to reflect upon the period of austerity in public (and defence) spending that immediately followed the collapse of the global financial system in 2008. Its effects are still being felt. Between the Strategic Defence and Security Reviews (SDSR) of 2010 and 2015, the national defence budget contracted by about 7.5 per cent, with whole capabilities removed from the UK order of battle. Air and Maritime platforms such as Harrier and Nimrod MRA4 were removed from service through decisions taken in SDSR 2010, so that the review itself was described by James Arbuthnot MP, at that time the chairman of the House of Commons Defence Select Committee, as:

> A clear example of the need for savings overriding the strategic security of the UK and the capability needs of the armed forces. The government needs to outline its plans to manage the gap left by the loss of these capabilities.[3]

This goes to the heart of the argument that capability, capacity, and underpinning competencies for the generation of national defence need investment from government. Moreover, that investment needs to be systemic, long-term, and assured. When the investment goes missing in action, as it did perhaps between 2010 and 2015, the competency pool contracts and the capability to generate national defence diminishes.

> It seems a legitimate concern that the UK government, of any political flavour, had failed to understand the causal, linear relationship between capability demanded and resources needed.[4]

This can be seen as follows: BAE Systems, the UK's largest defence business, in 2009 possessed the following characteristics:

- The business's direct value-added contribution to UK GDP was £3.3 billion.
- Productivity, as measured by economic value added per employee or full-time equivalent, was £78,175 as opposed to the UK national average of £42,200. This represents an 85 per cent delta uplift on the average value in favour of the defence business.

- The UK-registered company generated net exports of £4.8 billion.
- BAE Systems contributed £653 million in direct taxation to the UK government.
- Company direct research and development expenditure amounted to £900 million.
- The business sourced components, supplies, and services from other UK-registered companies, partnerships, and sole traders to the value of £4.1 billion.
- Within its supply chain, BAE Systems supported 125,000 UK jobs.[5]

As the UK defence budget shrunk from 2010 onwards, the management of BAE Systems responded by closing sites at Brough and Chadderton whilst, concurrently, announcing the end of shipbuilding at Portsmouth. The impact upon skills and competencies across defence would be profound.

A shift in the myth on competencies

Between 2010 and 2015, the authors undertook original research into a sample of BAE Systems employees who were exiting the business to take up other work opportunities or on redundancy terms associated with the company's rationalisation of its estate.[6] Assuming that our review during this period was representative of the broader defence enterprise in the UK, we concluded the following:

Competency leakage

The basic assumption often heard in government and across the military is that competencies and skills within the defence industrial base self-regulate through market forces. Consequently, when one UK defence company made headcount reductions another defence business absorbed these competencies. In this way they were said to be not lost to the UK.[7]

This has proved not to be the case. When workers have exited the defence ecosystem, over half of these people did so on either compulsory or voluntary redundancy terms. A similar proportion of people sought comparable work with another defence business, but only 20 per cent of them managed to find such an appointment. We can say, therefore, that 80 per cent of the personnel leaving a defence business is lost to the sector as a whole – their skills are not reabsorbed by another defence business. Moreover, 45 per cent of these exiting personnel are engineers, project managers, and information technology specialists; core skills deemed essential to the broader national defence effort.[8]

The case can be made, therefore, that without government sponsorship and intervention, defence skills do wither when left to free-market determinism. This has to be taken into account when defence and security strategies and policies are developed. That is not to say that the rationalisation and contraction of the UK defence industrial base should not take place – though we are not sure it is wise to deliberately reduce indigenous defence capabilities – but government should be aware of the impact of such policies on national skills and competencies.

Impact on tax revenues

This data suggests that government experiences both a short-term and a longer-term drop in revenues as a result of the redundancies made from the defence sector. The short-term drop relates to the periods people were out of work and thus not paying tax (they could also have been claiming job seekers allowances). The longer-term revenue falls relate to the number of people taking early retirement (in our study, 17 per cent), and those moving to jobs where they were less well paid (at least 40 per cent).[9]

Regional impacts

Significantly, just over a third of those leaving the defence company relocated to find other work. For areas where defence industry is important on a regional scale, such as Barrow, Preston-Blackburn, and Portsmouth, this suggests that defence cuts will have, potentially, a regional impact on local authority revenues, housing prices, and the returns to consumer-facing business as skilled people on good pay grades leave the local area. Indeed, a number of semi-structured interviews with respondents suggested that having to leave the local area to secure new employment was itself socially traumatic as well as economically impactful.

This short study is significant we would suggest. The work demonstrated that when a defence business makes headcount reductions, defence skills and competencies are not redistributed between similar businesses by a benevolent free-market. Rather, 80 per cent of personnel – and their core competencies – are lost to the defence sector. This is significant, as about half of the personnel leaving defence possess a scientific, technological, engineering, or mathematics background – subjects essential to a healthy defence sector and broader economy.

Moreover, government models addressing the impact of cutting the defence budget to drive public sector economies and commercial efficiencies should take into account likely redundancies from industry and the decrease in government revenues that would result. Defence savings to a significant extent hit government income, as well as reducing expenditure, and can have the effect of transferring state spending on an individual from a bounded and controlled defence contract offering specific defence benefits to, in the short-term, demand-led social security payments with little national benefit other than subsistence of the individual. If the government really wanted to secure economic and fiscal gain from cutting defence expenditure, it should make a particular effort when the economy is flourishing as opposed to when it is struggling or beginning to recover from recession.

It seems clear that redundancies from within the defence sector have a particular regional impact, with a third of personnel having to relocate to secure new employment. Defence cuts that are going to affect specific regions should be identified in advance by government and accompanied by national and European initiatives to boost high technology investment in those regions. In this way, the risks associated with a significant exodus of well-paid personnel, falls in house values, and so on can be mitigated and managed, leaving the region in question less exposed to economic uncertainties than would otherwise be the case.

Lastly, in imposing cuts on the on-shore defence industry, the government must recognise their impact on defence capability and the UK's capacity to develop, produce, and support equipment and services, and thus to enjoy the operational sovereignty that comes from not relying on external suppliers. Half the people leaving defence did not seek another job in the defence sector: this should ring a warning to policy makers and defence planners who value the UK's ability to respond to the threats of adversaries.

Competency thinking and defence reform

The development of skills and competency thinking in defence is intrinsically linked to broader defence reform, starting with the Smart Procurement initiative in 1998. The guardians of this initiative, subsequently rebranded as Smart Acquisition, were those people in government, the military, and industrial base who promoted and enacted its processes, behaviours, and objectives. The Acquisition Management System, a repository for acquisition tools and frameworks, stated quite unequivocally that Smart Acquisition placed a strategic emphasis on the development, training, and sustaining of people in acquisition through the development and implementation of competency frameworks.[10] Central to this commitment and investment was the development of the Acquisition Stream and Acquisition Leadership Development Scheme – personal development schemes that would roll-out these practices.

The Acquisition Stream was launched in February 2001 to create a stream of people in acquisition who were supposedly highly committed, skilled, and well-trained in defence acquisition and project management. Membership was voluntary and open to all military and civilian staff, as well as members of industry. The scheme operated through the development tools of an Acquisition Competence Framework, a personal development record, training, and development directory and development route-maps. Through these tools there was expressed a clear and robust methodology for working, behaving in the workplace, and developing one's career in a competency-based manner. This could be perceived as 'best-practice' in name, perhaps, but one-practice in design, roll-out, and execution.

The Acquisition Leadership Development Scheme operated for the perceived elite of the Ministry of Defence and defence industry as an extension of the Stream, and was intended to develop existing and future leaders in defence acquisition. The scheme was divided into three stages, foundation, core, and expert, with the primary differentiation being the competencies which an individual was expected to possess and the progress that they were said to have made against competency-based route-maps. The scheme was limited to 400 members, selected by competition against, once more, a pre-described competence framework. Smart Acquisition people were selected and developed against heavily prescribed requirements and procedures captured as competencies.

Within the acquisition function since 1998, the defence sector has favoured a mix of core competencies which describe the values, attitudes, and beliefs required from personnel, behavioural competencies that describe work-placed conduct and conventions, and functional competencies which describe the work-family, skills, and qualifications necessary for effective performance. A generic competence model is provided in Figure 7.1.

This functional competency model would supplement the core values, as follows.

The defence values for acquisition[11]

1 Recognise that people are the key to our success – equip them with the right skills, experience, and professional qualifications;
2 Recognise that the best can be the enemy of the very good – distinguish between must have and desirable;
3 Identify trade-offs between performance, time, and costs – cases for additional resources must offer realistic alternative solutions;
4 Never assume additional resources will be available – cost growth on one project can only mean less for others and for the frontline;

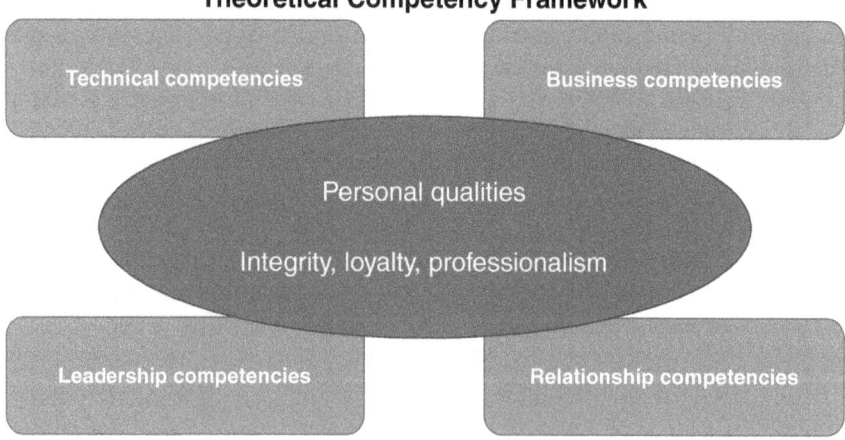

Theoretical Competency Framework

Technical competencies

Business competencies

Personal qualities

Integrity, loyalty, professionalism

Leadership competencies

Relationship competencies

Figure 7.1 Theoretical competency framework.
Source: the authors 2018.

5 Understand that time matters and slippage costs – through running on legacy equipment, extended project timescales, and damage to our reputation;

6 Think incrementally – seek out agile solutions with open architectures which permit 'plug and play'. Allow space for innovation and the application of best practice;

7 Quantify risk and reduce it by placing it where it can be managed most effectively – stopping a project ... can be a sign of maturity;

8 Recognise and respect the contribution made by industry – seek to share objectives, risks and rewards while recognising that different drivers apply;

9 Value openness and transparency – share future plans and priorities wherever possible to encourage focused investment and avoid wasted effort;

10 Embed a through-life culture in all planning and decision-making;

11 Value objectivity based on clear evidence rather than advocacy – ensure that we capture past experience and allow it to shape our future behaviour;

12 Realise that success and failure matter – we will hold people to account for their performance.

In many ways this reads like a generic list of managerial group-think, offering little concrete advice or direction for personnel to follow. In contrast, the acquisition behavioural competencies were clustered in six distinct groups relating to:[12]

• Applying a systems approach to a portfolio, programme, or project;
• Applying strategic direction;
• Initiating and managing change;
• Managing people and working in a team;
• Managing performance;
• Stakeholder management.

Within these six clusters a variety of behavioural competencies has featured through the years, supposedly guiding the actions of defence practitioners.

The armed forces: people and society

It is a management cliché that people are an organisation's greatest asset, and certainly in defence there is a need to think deeply and analytically about the roles and significance of people. Of course, there are some established ideas about people in the military, the civil service, and defence industries that we come on to discuss before looking at how these ideas are coming under pressure. What this tension represents for the defence enterprise as a whole is worthy of consideration.

Any discussion of the personnel dimension of the defence enterprise must take account of the politically-directed pressure for the armed forces to be broadly representative of the society they are meant to defend. In today's UK, that has implications for military organisational cultures and implies representation and opportunity for groups traditionally under-represented in the military, including women, ethnic minorities, and LGBT individuals and those who question their sexuality or are asexual.

Thought and practice in the military: some common perceptions

There are some established thoughts, perceptions, and practices relating to the military that are worthy of exploration:

1   In some areas of defence, there is a proper concern with attracting people in the right numbers. For many years, the Army has failed to recruit and retain enough individuals to meet its authorised strength, and there is an on-going specific challenge about attracting 35,000 people (in various categories) into the reserve force.

2   Certain parts of land forces still have need for people with limited intellectual capacity, but whose prime valued attributes involve emotional commitment coupled with strength, fitness, and stamina. A physical training programme of less than two years can do much to generate the latter attributes in any healthy young person.

3   However, the armed forces rely increasingly on equipment that requires a significant intellect, coupled with extensive training and education, for its effective use and support. The Navy and the Air Force, with their need for platforms just to operate in their respective environments, have traditionally focused on their need to ('optimally') man the equipment. Yet the Army also needs to move beyond a mind-set of 'equipping the man' given the increasing importance in its inventory of airborne platforms, surveillance and communications equipment, and advanced land vehicles.

4   The armed forces employ people in a wide range of 'occupations', some of which are closely related to occupations found in the civil sector (chefs and musicians), others are largely military specialised (aircraft armament fitters), and others/most lie in an interim position. 'Occupations' may be defined in terms of the educational and training 'qualifications' which they require and the 'experience' a post-holder is expected to bring.

5   Occupations can be classified according to the preparation and intellect they involve. An Australian defence analysis noted 'professionals', 'technician' grades, and 'trades'. Famously, Samuel Huntingdon defined the military officer as a 'professional' because of (his) specialist expertise in the use of lethal force.[13]

6   In terms of education, training, and experience, the military sector has traditionally 'grown' its own. That is, the military take people from their full-time education in school, college, and perhaps university, with the formal qualifications they have attained to date, and then develop and prepare people for military careers. The reasons for this are easy to identify: basically, the rigours of some areas of military life mean that young people are needed for many roles, and the specialist nature of some military

occupations means that the recruitment of someone prepared in the civil economy is not easily feasible: there are no individuals in the civil world who have the background to enable them to serve as the captains of nuclear submarines. For more senior command and leadership roles, it is widely believed that knowledge and experience of a wide range of military activities is needed. Huntingdon's assertion of the professional nature of the officer corps does much to explain and justify the 'grow your own' approach, with its implications that officers being groomed for senior roles must serve in a variety of posts while young, and that officers joining the military in their mid-20s have less chance of rising to the highest levels because they will not be judged to have attained a broad enough background in the time available.

Challenges to the status quo

There are a range of factors in the contemporary situation that challenge traditional thought about people in the military.

1   Technology means that the specific skills and attributes associated with some occupations are changing: directing a modern artillery piece requires keyboard skills rather than the strength rapidly to turn a handle for a sustained period. The Bowman communications system places more demands on its operators than did its Clansman predecessor. The occupation of aircraft navigator has disappeared.
2   Technology is generating new occupations for which the standards have yet to be generated. Drone pilots and 'cyber warriors' are cases in point.
3   Technology is generating some cases of occupations that require costly and long-duration training that, in the case of officers, hinders their chances of generating the range of experience defined as needed for promotion to very senior ranks: Apache pilots are a case in point.
4   The military has already included a few specialists who need an extended preparation period and are employable in the private world. Such people (doctors, vets, and lawyers most obviously) are given the military rank needed to provide them with the necessary pay level. The range of such specialists could grow, with the US Navy already looking to recruit cyber experts directly into non-commissioned and officer grades.
5   The support needs of complex equipment continue to grow, one of the contributing factors to debates about the support tasks that should be undertaken within the government machine and those that should be entrusted to the private sector.
6   Arguably we are living in an age of 'disruptive technology' as far as the UK military is concerned, which is changing the relative importance of organisational sub-units within the military and of the personnel who operate in them. Given the growing number of vital activities occurring far from the geographical 'frontline' (if such a location exists), will it remain persuasive

that the higher levels of armed forces command should be entrusted to those who have experience of putting their own lives at significant risk? The extraordinary arrangements being made to recruit military reservists with cyber expertise is a manifestation of established military culture being challenged to accommodate people with novel but valued expertise.

7   Given the formal roles of the military with regard to military aid to the civil power, it would appear logical that the officer corps need expertise in a wider syllabus than the effective use of lethal force, including disaster management, and working with other organisations and cultures.

The pressures associated with these considerations are having visible effects, not least in the area of the generation of cyber capability.

From the military to the defence enterprise

But the military represent only the most visible aspects of defence capability: they are dependent on an extensive supply of goods and services from non-uniformed public-sector workers and the private sector for their capacity to sustain current activities. They also rely heavily on the private sector for the innovation and development work necessary to generate the innovative equipment needed for future operational advantage and success.

More than half the UK defence budget appears to be spent with the private sector and more than half of the people 'working in defence' are employed outside government. Viewing defence as an extended enterprise generates attention to the people in the present and future supply base of the armed forces.

We calculate that within the supply base there are 142,000 direct employees servicing the needs of the military – of these, over 4,000 are apprentices. Moreover, there are an additional 400,000 jobs that indirectly support defence businesses in the UK alone, with many of these people employed in the 5,000 small-to-medium sized enterprises that support the defence sector.[14] These people are typically organised into the 'job families' shown in Box 7.1 across industry and commerce:

There are, therefore, in broad terms, 18 job families or roles that relate to the defence sector in the private sector. In contrast, there are 53 roles in the Royal Navy, 124 job types within the British Army and 64 roles advertised for the RAF: 241 roles all with their own competency frameworks, recruitment and selection criteria, and training regimes.[15] This is of course supplemented by a further 65 roles relating to defence that are offered through the Civil Service. The defence environment in the UK presents as a complicated and bureaucratic place when viewed through the lens of competencies.

New Public Management

The influence of New Public Management thought, with its stress on the effectiveness of competitive markets and customer structures for the efficient

---

Box 7.1 Job families

Corporate Executive/Director
Programme/Project Director
Engineer – Chartered
Scientist
Research and Development
Production Manager
Project Management
Engineer – Construction/Fitting
Information Technology
Supply Chain/Logistics
Support – Technical
Support – Manual/Maintenance
Business Development/Sales/Marketing
Human Resources
Training
Finance
Design
Health and Safety

Source: authors' analysis of defence job clusters within BAE Systems

---

generation of goods and services, and its subsequent lack of faith in public sector bodies to generate efficiency, has contributed much to the outsourcing tendency in UK defence. Today, back-office functions, facilities management, education and event training, and many aspects of equipment support, are among the defence functions generated by the private sector.[16]

The people management dimension of the DEE depends on the importance attached to national private sector capabilities in the innovation/development/production and support domains, the perceived role of international collaborative activities, and the readiness to rely on other governments to manage relevant issues. It is, in short, a matter for risk management.

In the MoD, aside from sharing the national concern with the generation of graduates and technicians with Science, Technology, Engineering, and Mathematics, there has been little concern about expertise in the broader defence sector. The priority has been to expose UK suppliers to competition from outside the UK and regularly to accept dependence on a foreign supplier for equipment, spares, and other elements of support. Buying from a foreign production line presents comparatively few time or finance risks in the short term compared with national or collaborative development projects.

There have been some exceptions to this practice, in particular:

In the nuclear submarine, nuclear power generation and nuclear weapons fields, where competition for those with nuclear qualifications is intensifying between the commercial and defence sectors, there have been parliamentary and government studies into nuclear expertise availability and development in the UK.

The Team Complex Weapons construct dating from the Defence Industrial Strategy of 2005 has involved the MoD and industry working together to enable MBDA and its partners to maintain a workforce that can design, develop, produce and support a range of missiles relevant to the UK's needs.

A capacity to design and build conventional warships has also been singled out for attention, with recognition that a drumbeat of orders is needed for a comprehensive range of workforce capabilities to be sustained.

The position in the CBW field is more opaque, with the UK efforts being research-focused.[17]

Particularly when it comes to the capability to develop new systems, it is not enough to aggregate the STEM-related skills of the individuals concerned: the managerial and collective capabilities of the organisation also form a vital element. Organisations that do not get a chance to practice these capabilities see them fade. Australia is a good case study to consider, as shown in Box 7.2.

That a Defence Ministry, as the UK's prime contractor for defence capability, should be concerned with the present and future state of the human resources in its present and future supply chain would not seem extraordinary to a large manufacturing company. It is normal for such firms to pay significant attention to the human resources in their supply and support chains, and to understanding the risks and opportunities involved.

Impact on the defence enterprise in peacetime and on operations

A final set of people considerations relate to basic characteristics of UK defence. Except for the funding of those charged with a continuous deterrence role, the greater part of the defence budget is for the preparation of capability available for use at government discretion. For all of the twenty-first century and most of the Cold War period, UK forces have been significantly occupied on operations, which has required distinct extra funding. Put simply, the full-time armed forces

---

Box 7.2   Australia

Australia is a case of a government that seeks to maximise involvement of the national defence industry in the support of its equipment and it is thus much concerned with monitoring and promoting the capabilities of the workforce in its defence firms. For Australia, the training and experience that the military can provide to its personnel who later go to work in industry is a source of assurance, because it should reduce the pressures on the private sector to recruit, develop and retain personnel. Its document does not indicate any sense of 'theft', that industry 'steals' our military personnel.

of the UK are paid largely to do nothing in particular, but to be ready to undertake a range of missions.

From a financial perspective, in times of calm it makes sense to minimise those resources maintained on a full-time basis and to maximise less expensive 'part-time staff' (reservists) where possible. It also makes sense to make use of the commercial civil economy to purchase services, not least for the support of troops but also for transport, that are not required in peacetime but for which there is a demand when a large sustained military operation is underway. In the logistics world, it is accepted that the MoD need not hold large stocks of items, especially those with a limited shelf-life, which would be needed for an operation but are not needed in peacetime on the same scale: small batteries for personal devices are the obvious example. The same logic can be and is applied to people: the MoD does not employ on a full-time, constant basis, those who can be found in a suitably qualified form in the civil economy.

It also can make sense for the military to put some of its full-time personnel to work in the civil economy/civil sector, not least when that gives them the chance to maintain and develop their skills. This happens with military medical personnel who are put to work in the National Health Service.

How the defence enterprise is to define the optimum balance of full-time personnel (who should need extensive and constant training), reservists of various descriptions, and those who will be hired on a contract basis solely for the execution of a specific operation, is a major planning challenge.

The competency-based system

As New Public Management and the rigours of an ever-expanding scientific and programme management agenda has impacted upon the defence environment, it seems reasonable and timely to remind the reader that this has generated a complex system of management that goes beyond competencies: 'There are multiple levels and types of competencies across defence and no one person really understands how it all comes together as a coherent system. Perhaps somebody needs to?' [18] This can be understood especially when we consider the range and depth of Services, ranks, businesses, levels of seniority, management layers, skills, and values residing in the public and private entities that comprise the defence sector.

Across businesses, within the armed forces, and throughout the civil service the use of stated competencies in target-setting for organisations, teams and individuals, and the subsequent systematised performance management of these entities, is one of the principal growth areas in organisational development since 1991.[19] This has spawned the mushrooming of competency frameworks across defence, such as the acquisition competency framework mentioned above and the Framework for Nuclear Engineering Competencies[20] which we come on to discuss shortly. As a consequence, there is hardly a person working in defence, in business, or the public sector, who is not located in one or multiple competency frameworks. These frameworks themselves can be grouped into a series of job families, perhaps adding to the sense of a competency-based enterprise structure, but also to rising concerns around complexity and responsiveness. For rigid competency regimes seldom aid organisational agility.

This seems an important point. The defence enterprise is there to protect the British people from multiple threats, both state-based and those emerging from non-state actors. It has to be responsive and focused. Competency-based frameworks assume that defence management is scientific and causal, whereas management, for many people, is an art, driven by context not code. Competency frameworks are often rules and objects of control – the very notion of a performance management system is a control mechanism. Defence people seem often to prefer clear direction and delegation of responsibility rather than the imposition of rigid rules. The critique of the competency-based system in defence is that principles might actually be better, systemically and psychologically, than rules.

A way to illustrate this is to think of inventory management within organisations. A principle of lean manufacturing is that inventory is kept to a minimum commensurate with the daily flow of construction or assembly. To hold more inventory than needed is to tie-up valuable money, people, and other resources in managing and securing stocks.

> Rules can also be seen as inventory. The static, dictatorial nature of rules means that you are constantly spending money enforcing them and arguing for their existence. People need to be put in charge of the rules and those people become mired down in their adjudication.[21]

If it could be argued that principles were better for defence than a rules-based competency system, what principles should apply? Of course, there is no definitive way to answer such a question but the performance literature does seem to point to the following themes:

- The importance of planning for activities and throughput rather than outcomes or effects;
- The centrality of treating people as professionals and to encourage freedom of association, creativity, and team problem solving;
- The importance of respect and empowerment of decision-making throughout the enterprise;
- The permission for processes to reform continually and the encouragement of emergent design and challenge;
- The sense that clarity drives improvement and responsiveness, reduces waste, and delivers sought organisational outcomes.

The quest to bring values and principles into frameworks is at the forefront of organisational and personnel development initiatives. Yet this tension between competence/rules/performance management and freedom of action/creativity/permission, which can be seen in most complex systems, is not a stranger to defence. Its effective management is critical in keeping the nation-state safe and secure.

Nuclear engineering competency framework – an example

As an example of this thinking, the development of a framework for nuclear skills and competencies across UK defence was undertaken from 2016 onwards, consisting of four domains: the Core, Behavioural, Functional, and the Local. It was considered that skills would feature within the Functional area of a comprehensive public–private Competency Framework. The development team drawn from government and industry recognised that significant work was required to embed, as common practice, the working methods and assurance regimes from myriad businesses and government teams involving more than just a focus on engineering and systems competencies. However, an initial focus on nuclear skills was considered appropriate, with other elements of a comprehensive framework to be developed in slower time.

A detailed analysis was undertaken of the existing, and sporadic, organisational competency frameworks provided by organisations such as the MoD, Babcock Marine, Rolls-Royce Submarines, BAE Submarines, and the Royal Navy, the latter deliberately separated from the MoD competency set. A tool combining the key elements of all of these approaches was developed and labelled as the Defence Enterprise Nuclear Skills Conceptual Competence Framework-Comparison Tool (DENSCCF-CT). The tool included a set of competence statements formed by amalgamation and alignment (where possible) of statements drawn from each of the contributing frameworks. For alignment purposes, the MoD Nuclear Competence Framework was taken as the baseline, since this was determined to have the most comprehensive statements and standards. The MoD's framework covers:

- Safety and Security, Defence Nuclear Programme Management;
- Emergency Response;
- Concept, Assessment and Design of Systems and Infrastructure;
- Manufacturing, Testing, Commissioning and Acceptance;
- In-Service Support and Maintenance;
- Liabilities Management (Decommissioning and Disposal).

Interestingly, 49 per cent of competence statements from the other organisations were found to align with the MoD Framework. This level of alignment might seem odd given that nuclear engineering is an area so heavily regulated by safety standards, but hints at the difficulties associated with the development of comprehensive competency frameworks. As a senior defence figure commented: 'A lot of effort went into this thing but I wonder if we knew why, other than the fashion for generating competence frameworks.'[22] Nonetheless, as well as providing an integrated framework of competence statements, the DENSCCF-CT provided an analysis against which the competence statements of roles and responsibilities submitted by partner organisations could be assessed and judged. Also, the tool yielded a set of nuclear competences gleaned from frameworks currently utilised across the whole defence enterprise and, in an era of the managerial, such an outcome, self-referencing though it may be, for many is worth the effort.

Competency frameworks and performance management

Competency frameworks are linked to the selection and development of individuals and teams, and the inevitable measurement and assessment of performance. Within this context: 'It is not too difficult to conceive how performance management should function. It is much harder to ensure that it works in practice. It takes time, energy and determination.[23] The supposed need for a performance management regime, based on individual, team, and organisational competencies, includes the following:

- The need to reinforce a performance culture or to help change an existing culture to one that embraces performance;
- The ambition to weld-together different organisational entities into a singular culture;
- The desire to improve the performance of individuals and teams, or at the least the assessment of such;
- A commitment to develop the potential of employees;
- A perceived need to provide the information and measurement tools required for reward, remuneration, and promotion;
- A need to integrate organisational, team, functional, and individual performance objectives and targets;
- The wish to provide a framework within which managers can assess performance and staff can understand judgements around work.

This is not limited to the defence sector, of course, but as the defence enterprise is informed from broader society, so the need to satisfy these drivers of performance form in the minds of defence decision-makers. As a senior defence leader states: 'We want to learn what we can from other sectors and practices and apply it to defence.'[24] Accordingly, across the branches of the armed forces and within the business that populate the defence enterprise, a highly formal, centralised, and hierarchical cats-cradle of competency-based performance management regimes exist, constantly measuring team and individual performance. Personnel management and performance management are synonymous in this regard, isolated perhaps from broader judgements around operational effectiveness and the ability to generate battle-winning capabilities.

Conclusion

Within this chapter we have sought to explore and explain the centrality of skills and competence-based thinking across defence. Naturally, the evolution of that thinking relates to ongoing and continuous defence reform, itself informed by changes in thinking relating to organisational development, people management, and performance. The defence enterprise does not sit isolated but skips to the tunes played in broader society.

Competency, for our purposes, has been defined as a capacity to deliver a particular service for a stated, pre-planned effect through the application of skills and knowledge, attested through qualification or certification. By its very nature this is systemic and comprehensive: given the complexities of the defence enterprise there is little wonder that an industry has evolved focusing on competency-based management and the measurement of performance.

This thinking, though, has generated one or two assumptions that are seldom questioned in the sector. The need for competency frameworks and management are never challenged – to do so would leave the challenger exposed to accusation of being a modern apostate, ignorant of critical working and managerial practices. Second, in defence, the common assumption is that when businesses, especially, release staff through redundancy or efficiency programmes, through the dynamics of competency management and the hidden hand of a free-market economy, these defence people and skills are absorbed elsewhere across the sector. Our work shows that this is not the case. Rather, 80 per cent of personnel leaving a defence business are lost to the sector as a whole and those skills are not replaced – capability within defence contracts rather than self-regulates.

Moreover, as we have seen, a competency system – albeit one anchored in the language of reform and continuous improvement – suffers from the inevitable rigid and absolutist structure associated with its construct. For those who believe that defence and security require themes of agility, flexibility, and responsiveness, competence thinking can prove challenging indeed. As a consequence, a sense of the wellbeing of the defence enterprise through a competency review seems meaningless. Rather, what needs to be undertaken is a comprehensive audit of the defined occupations, roles, and populations within the current force structure (government, military – regular and reserves, and industry) in terms of trades, technicians and professionals. This could be tuned to their professional grade or rank, and the formal qualifications from education, training courses, and time in role required to generate a suitably qualified and experienced person for that occupation. But, then, there are close to 250 separate roles in the military and 18 job families in industry. Such an audit would be a mammoth task. Perhaps we have constructed a cats-cradle of overlapping competencies and roles that defies review as a system. That would be a very disturbing finding for citizens and MoD leaders alike.

Notes

1 HM Government, Defence White Paper, *Defence Industrial Strategy* CM6697 (London: The Stationery Office, 2005), para A4. 22.
2 Interview with Lord Arbuthnot, a retired UK defence minister, on 7 March 2017.
3 House of Commons Defence Select Committee Announcement, 1 August 2011, 'Wide-ranging Concern about the Strategic Defence and Security Review' (London: HOC, 2011).

4 See John Louth, Trevor Taylor and Henrik Heidenkamp, 'Defence Skills: A Shift in the Myth,' *RUSI Briefing Paper*, 2014.
5 Oxford Economics, *The Economic Contribution of BAE Systems to the UK in 2009* (Oxford: Abbey House, 2011).
6 See Louth, *et al.* (2014), op. cit.
7 See Jacques S Gansler, *Democracy's Arsenal: Creating a Twenty-First-Century Defense Industry* (Cambridge, Massachusetts: The MIT Press, 2011).
8 See, for example, the collection of papers edited by David Moore, in: David Moore (ed.), *Case Studies in Defence procurement and Logistics: Volume 1 – From World War II to the Post Cold-War World* (Cambridge: The Cambridge Academic, 2011).
9 See Trevor Taylor and John Louth, 'The Destinations of the Defence Pound,' *RUSI Briefing Paper*, 2012.
10 Ministry of Defence, *The Smart Acquisition Handbook* (Edition 4), (London: MoD, 2002).
11 Ministry of Defence, *Defence Acquisition* (Edition 1), (London: MoD, 2008) p. vi.
12 Ministry of Defence, *The Smart Acquisition Handbook* (Edition 5), (London: MoD, 2004).
13 Samuel P Huntingdon, *The Soldier and the State: The theory and politics of civil-military relations* (London: Harvard University Press, 1957).
14 See ADS, *2016 Industry Facts and Figures: A Guide to the UK's Aerospace, Defence, Security and Space Sectors* (London: ADS, 2016).
15 www.mod.uk, accessed on 30 March 2017.
16 See Ilan Oshri, Julia Kotlarsky and Leslie P Willcocks, *The Handbook of Outsourcing and Offshoring* (Edition 3), (London: Palgrave Macmillan, 2015).
17 During the 1980s, it is understood that a secret study was carried out as part of the early Levene Reforms to emphasise competition in defence procurement. Anecdotally it is understood that a few areas were identified where foreign suppliers would never sell the UK their most advanced technology, if they would sell that technology at all. It was therefore concluded that the UK needed to maintain national capabilities in these areas.
18 General Sir Nicholas Carter, Chief of the General Staff, speaking to the authors in March 2016.
19 See Michael Armstrong and Angela Baron, *Performance Management: The New Realities* (London: CIPD, 2000).
20 Defence Enterprise Nuclear Skills Conceptual Competence Framework-Comparison Tool (DENSCCF-CT), developed by the MoD and industry in 2016.
21 Maritza van de Heuvel, Joanne Ho and Jim Benson, *Beyond Agile: Tales of Continuous Improvement* (Seattle: Modus Cooperandi Press, 2013) p. xix.
22 Private interview with senior official within the MoD, June 2017.
23 See Michael Armstrong and Angela Baron, *Performance Management: The New Realities* (London: CIPD, 2000) p. 357.
24 Air Chief Marshal Sir Stuart Peach, Chief of the Defence Staff, speaking at RUSI on 14 December 2016.

Bibliography

ADS, *2016 Industry Facts and Figures: A Guide to the UK's Aerospace, Defence, Security and Space Sectors* (London: ADS, 2016).

Armstrong, M and Baron, A (2000), *Performance Management: The New Realities* (London: CIPD).

Gansler, JS (2011), *Democracy's Arsenal: Creating a Twenty-First-Century Defense Industry* (Cambridge, Massachusetts: The MIT Press).

van de Heuvel, M, Ho, J and Benson (2013), *Beyond Agile: Tales of Continuous Improvement* (Seattle: Modus Cooperandi Press).

HM Government (2005), Defence White Paper, *Defence Industrial Strategy* CM6697 (London: The Stationery Office).

House of Commons Defence Select Committee Announcement, 1 August 2011, 'Wide-ranging Concern about the Strategic Defence and Security Review' (London: HOC).

Huntingdon, SP (1957), *The Soldier and the State: The theory and politics of civil-military relations* (London: Harvard University Press).

Louth, J, Taylor, T, and Heidenkamp, H (2014), 'Defence Skills: A Shift in the Myth,' *RUSI Briefing Paper*.

Ministry of Defence (2002), *The Smart Acquisition Handbook* (Edition 4), (London: MoD).

Ministry of Defence (2004), *The Smart Acquisition Handbook* (Edition 5), (London: MoD).

Ministry of Defence (2008), *Defence Acquisition* (Edition 1), (London: MoD).

Moore, D (ed, 2011), *Case Studies in Defence procurement and Logistics: Volume 1 – From World War II to the Post Cold-War World* (Cambridge: The Cambridge Academic).

Oshri, I, Kotlarsky, J, and Willcocks, LP (2015), *The Handbook of Outsourcing and Off-shoring* (Edition 3), (London: Palgrave Macmillan).

Oxford Economics (2011), *The Economic Contribution of BAE Systems to the UK in 2009* (Oxford: Abbey House).

Peach, S, Air Chief Marshal, Chief of the Defence Staff, Speech at RUSI on 14 December 2016.

Taylor, T and Louth, J (2012), 'The Destinations of the Defence Pound,' *RUSI Briefing Paper*.

www.mod.uk, accessed on 30 March 2017

8   Defence as community action

Introduction

This chapter deals with the importance of wider national society for the professional defence sector, why the relationship can be problematic, and how the government is seeking to manage the situation, not least through the mechanisms of the Military Covenant. It begins with four basic areas of challenge for defence as regards its relationship with wider society.

Chapter objectives

By the end of this chapter the reader will understand:

1   The complexity of issues facing the military within society;
2   Veteran employment issues and the skills challenge;
3   The military on operations and the support offered within the community;
4   The Military Covenant – its purposes and implementation;
5   The Whole Force and its implications for the military-community relationship.

Chapter structure

This chapter starts with a review of what we describe as the 'problem space' for the defence community within broader society, before considering employment and employability of veterans. We go on to consider the roles played by the community when military personnel are on operations. This leads us to a consideration of the UK Armed Forces Covenant, its implementation since 2010, and its effect. We then conclude with the challenges posed by the 'Whole Force.'

Defence and society: the problem space

Demonstrating and perceiving the benefits of defence

From the very emergence of the sovereign state, indeed arguably justifying its existence,[1] defence has been a public service for which the state has been responsible. Adam Smith most famously articulated this and signalled that the execution of this responsibility was not straightforward.

The first duty of the sovereign – that of protecting the society from the violence and invasion of other independent societies – can be performed only by means of a military force. This may be effected either by obliging all the citizens of the military age, or a certain number of them, to join in some measure the trade of a soldier to whatever other trade or profession they may happen to carry on; or by maintaining a certain number of citizens in the constant practice of military exercises, thus rendering the soldier's occupation a special profession, distinct from all others. A militia is the less expensive, but a standing army is by far the more efficient, defence; and its cost falls to be borne by the sovereign or the commonwealth.[2]

The art of war, however, as it is certainly the noblest of all arts, so in the progress of improvement it necessarily becomes one of the most complicated among them. The state of the mechanical, as well as of some other arts, with which it is necessarily connected, determines the degree of perfection to which it is capable of being carried at any particular time. But, in order to carry it to this degree of perfection, it is necessary that it should become the sole or principal occupation of a particular class of citizens, and the division of labour is as necessary for the improvement of this, as of every other art. Into other arts the division of labour is naturally introduced by the prudence of individuals, who find that they promote their private interest better by confining themselves to a particular trade than by exercising a great number. But it is the wisdom of the state only which can render the trade of a soldier a particular trade separate and distinct from all others.[3]

However, the existence of incontestable links between military spending and peaceful consequences are usually difficult to demonstrate with no counter factual being available. The UK has spent very substantial sums on the development and continuous deployment of a nuclear deterrent force which politicians regularly assert has kept the UK safe. But the sceptical could argue that the absence of a conventional attack on the UK homeland or that of NATO allies had nothing to do with the UK deterrent and might reasonably ask in which particular crisis did the UK deterrent play a key role? Spending on health or education can be targeted to achieve specific desired outcomes not currently being achieved, such as reduced waiting times for cancer treatments. But linking spending efforts and outcomes is a much more demanding task for the defence sector where there is often uncertainty about the real intentions of perceived adversaries and whether and how they have been modified by another country's behaviour.

The roles and therefore powers of government in sovereign states, especially in the more economically advanced societies, have expanded significantly in the past 200 years. Transport, education, health, the environment, macro-economic management and regulation, agriculture, and even culture are today all matters for government, to the discontent of some who are significantly opposed to the idea of 'big government'. A feature of these many government activities is that they directly impinge on people's lives in ways that defence does not. The population does not directly experience successful defence in the way that it does health, education, transport, and many other services. Virtually all citizens go to schools, on occasions visit the doctor, and use the roads and rails. Some even go to galleries and museums. But hardly anybody has any personal beneficial experience with defence, even if they are lifted by helicopter off Mount Snowden in winter (this service has been privatised). When defence is working well, the population tends not even to notice. Defence can thus be easily overlooked in the public's consciousness, at least until things go wrong. Joseph Nye's insight from 1995 tersely captures the situation. 'Security is like oxygen – you tend not to notice it until you begin to lose it, but once that occurs there is nothing else that you will think about.'[4] Thus, a constant challenge for governments is to persuade taxpayers of the good sense of paying for something whose benefits they rarely experience and which anyway are difficult to demonstrate.

The British armed forces as small, isolated, and mysterious

A further consideration is the contemporary small size of the UK defence effort and the lack of everyday societal contact with defence forces.

The British armed forces are a small and decreasing element in the UK population. To grasp the scale of change, it is not necessary to consider the mobilisation of the whole of society during the Second World War alone, when the UK had almost five million people in uniform, or to go back to the days of conscription (which was phased out from 1957), when the armed forces numbered around 700,000.

At the end of the Cold War, there were 305,800 people serving full-time in the Army, Navy, Air Force, and Marines, from a population of 57.2 million: i.e. the armed forces were about 0.5 per cent of the population.

In December 2016 the strength of the armed forces was 197,150 out of a population of 64.7 million, i.e. the armed forces' share of the population had fallen to 0.3 per cent. These armed forces constituted just 0.57 per cent of the UK working population.

These numbers mean that fewer and fewer people have experience of serving in the armed forces, or have relatives and acquaintances who have served in the armed forces. Add in that defence is widely perceived as a technical, complex, and difficult to understand set of activities, the foreseeable result is that the number of people feeling they have any understanding of the defence sector is small. Our personal if anecdotal experience is that even people with close relatives in the armed forces often have little idea of their particular role.

Thus, a challenge for the Ministry of Defence is to get people to support with their taxes an institution of which they have little understanding and with which they have little or no contact.

This has to be linked to the relative isolation and lack of visibility of the armed forces. The MoD is the largest landowner in the UK but, for reasons of economy and to reduce the number of house moves for its people, has long been seeking to focus the stationing of the armed forces at fewer places. The Army is increasingly based around Aldershot in the south, in Andover and eastwards around Salisbury Plain, and around Catterick in north Yorkshire. The Navy is focused on Plymouth and Portsmouth, and to a lesser extent around Faslane in Scotland. The RAF, for historical, military, and then environmental reasons, is mostly in the less populated areas of Lincolnshire, with all the transport aircraft now based at Brize Norton in Gloucestershire. For security reasons, all MoD installations have to be surrounded by prominent fences and protected by guards. The Irish troubles meant for decades that military personnel did not wear their uniforms in public, and Islamic terrorist threats have caused this practice to continue.

There are exceptions to this relative isolation, notably when the armed forces are called on to undertake non-military functions in support of the civil authorities: after the turn of the twenty-first century, the armed forces helped with dealing with the foot and mouth outbreak of 2001 and with flooding emergencies. These involved only a very small proportion of the UK population and the most prominent case of UK forces interacting with the wider population occurred in 2012, when they acted at short notice to deal with security at Olympic venues. When the military are called on to support the civil power, i.e. when the police are unable to cope with a situation involving violence, it is difficult for the military to sustain a positive image, certainly among at least one side in a conflict. The British Army's experience in Northern Ireland after 1969 underlines this point.

But activities involving support to the civil power and to the civil authorities are exceptional: the prevailing general picture is the military for the public as a whole is an invisible mystery.[5]

This has implications for those supposed to hold the governmental sector to account in defence, most obviously parliament and the media. Members of parliament often take little interest in defence unless things are clearly going wrong because few of their constituents are actively concerned: developments remote from the ordinary person's life, that are anyway hard to understand, tend not to loom large in individuals' consciousness. Of course, defence-related employment considerations in MPs' constituencies do grab their attention, and parliamentarians from areas that rely significantly on defence, such as Barrow, Portsmouth, Preston, and of course the Clyde, must understand at least chunks of the defence agenda.

The situation and attitudes of those working in defence industry in the UK could be dealt with either in a chapter on defence and the community or the chapter covering defence acquisition. Such people are clearly a significant element of the Defence Extended Enterprise,[6] but their drivers and attitudes are little explored. In some original in-depth qualitative research undertaken by RUSI colleagues in Barrow, a nuanced picture emerged in which many staff were influenced by more than transactional considerations of money and job security. A common sense of pride linked to a wider communal identity emerged from the production of such demanding and important products (nuclear submarines). There was considerable consciousness that many sailors would entrust their lives to the platforms being built and of the importance of the boats for national security.

In addition, the small size of the uniformed forces means that not many MPs have direct experience of defence in their past employment: before the 2017 election there were about 50 out of 650 members of parliament with some military experience, including those who were still reservists, including the Armed Forces Minister in 2017, Mark Lancaster.[7] While over 8 per cent of the total was a higher number than might have been expected, it has to be put alongside the 100 per cent of MPs with experience of the health service.

The Armed Forces Parliamentary Scheme, under which MPs can learn about defence and even spend a week with a particular military unit, represents a modest effort to mitigate this concern, but the take-up by parliamentarians is on a small scale.

The House of Commons Defence Committee, ably supported by academics and retired military, as well as the National Audit Office, has some enthusiastic and informed members, and scrutiny of its reports reveals options to study many significant questions and findings that reflect awareness of the complexities and risks of defence. However, defence expertise is not a feature of all its members, as some will admit.

The broadcast and print media are the main channels of communication between the governmental defence sector and British society. All the 'broadsheet' newspapers maintain a small team of security and defence reporters and commentators, and the MoD has a communications department headed by a two-star civil servant, which is part of the government communications service. The latter's website proclaims that 'Our aim is to deliver world-class communications that support Ministers' priorities, improve people's lives and enable the effective operation of our public service. We serve both politicians and the public'.[8] In practice, reconciling these aims in the defence sector, as elsewhere, is often not straightforward given that ministers often feel an interest in controlling the dissemination of negative news.

Moreover, because defence is a complex field, the significance and even nature of 'facts' is often uncertain: take for instance a claim in *The Times* newspaper in July 2017 that the price per aircraft of the F.35B to the UK was £150 million. Those familiar with combat aircraft costs could reasonably respond that it might be, or on the other hand it might be less, or a lot more. As of the date of the article, the UK had ordered 14 aircraft while intending eventually to buy 138. If production of the F.35 family is stepped up as envisaged, actual manufacturing costs will continue to fall so future 'flyaway' costs should be less. However, a modern combat aircraft, like all platforms, needs not just an initial supply of spares but also in excess of 300 items of test equipment for the components and sub-systems on board. All these items are needed if the fleet is 12 or 50. Because of flying costs (cited by *The Times* at £54,000 per hour), advanced simulators are needed so that the actual aircraft can be used less. The F.35B exists only in a single seat version so a simulator is needed to train the new pilot before it is safe for her or him to take an actual aircraft into the air. Then there are infrastructure costs associated with the F.35, including the provision of security structures to protect the aircraft but, more obviously, the right hangars and landing surfaces are needed. In June 2017, the MoD announced a contract worth £135 million for a hangar and other infrastructure work at Marham for 12 aircraft, which in itself represents more than £11 million per aircraft.[9] Is it reasonable to add the UK share of the R&D costs (over £2 billion) of the aircraft to any cost per aircraft calculation? These costs are all associated with tangible items and do not include training costs for operators and maintainers. While it is certainly true that the flyaway/production cost of the aircraft is only the start of expenditure, a precise cost per aircraft will vary massively with the assumptions made about the other elements of expense. However, this is not the kind of detail that even the broadsheet newspapers in the UK are likely to want to explain, and anyway the MoD may not always wish to disclose.

A related consideration is that, compared with other government departments, defence spends money in big chunks. At 2 per cent of GDP, defence represents a sixth of spending on welfare and pensions, a third of expenditure on the NHS and almost half what is spent on education. Popular awareness of these ratios is believed to be modest, but defence is alone in not delegating much of its spending to the regional and local levels and in having many items of spending that are measured in the millions. When things go wrong, the numbers are also in the millions and, on a quiet news day, unsuccessful expenditure of a few millions can justify a headline.[10]

Over time, relations between the press and the MoD have ebbed and flowed in their quality. Some predictable patterns can be expected: journalists and outlets that publicise only negative news about defence tend to fall out of favour. If the MoD is tempted to deny access to specific journalists who have exposed negative information, it does not incentivise the media to allocate good people to the defence sector or even to cover defence at all. The relationship is not and cannot be easy, given the media's need to report things to which the public will pay attention, and of course the necessary need for keeping some information secret on security grounds.

Clearly the security considerations play a significant role in the MoD's capacity to communicate with the public. On the one hand, the government recognises the need for taxpayers to know how their defence money is being spent and that they are getting a return in terms of capability. On the other hand, there are facts that the MoD wishes not to release for security purposes. However, to this must be added the consideration that the release of some information might be damaging for the government or a specific minister, and civil servants are there to support the elected government. The precise content of annual MoD documents, such the Annual Report, Accounts and the Equipment Plan, and the statistics about defence released by the Office of National Statistics, reflects the MoD's efforts to manoeuvre through these considerations.[11] The experience of the authors is that overall the British Ministry of Defence is less open than its equivalent in the United States and more transparent than the German Verteidigungsministerium!

To summarise, defence needs public and parliamentary support to secure needed resources, yet defence is difficult to understand and not at all central to the daily lives of the great majority of the British people. The defence establishment wants to control the release of information in order to sustain its relations with wider society, whereas the media is concerned with the dramatic story and is conscious that bad news tends to command more attention than routine success. Thus, the MoD's relations with wider society cannot be managed according to a simple formula, but require constant judgement and attention.

The military veteran as employee in the civil sector

A further consideration is that, for most people who serve in the military, it is not employment for the whole of their working lives: they want and often need meaningful work after they leave the armed forces. The British armed forces comprise volunteer professionals and the physical demands of many of the roles make it highly desirable that many of them are 'young', i.e. under 35. Thus, in an era when the end of men's working lives is expected to be well over 65, most people joining the military do not seek lifetime employment, and anyway the military do not want them for anything like that period. Most officers are expected to retire at 55 at the latest and enlisted personnel at 48.

For the armed forces to recruit and retain for a decade or more the talent they want, it is highly desirable that an individual's experience in the armed forces should wherever possible support their future civilian employment prospects. This requires both that they should possess skills and attributes valued in the commercial and wider civil sector, and that the civilian world should perceive accurately the benefits of employing those with a military background.

Going back decades, the military tended to train and educate its people with qualifications that had no direct civilian equivalents, thus helping tie them into military careers. It then changed direction so that most military qualifications matched the national qualification system. The MoD recognises that a healthy professional armed force needs to develop people who can enjoy meaningful and rewarding work after the end of military service. But the transformation in qualifications is not complete. While Masters degrees are available from both the Advanced Command and Staff and the Royal College of Defence Studies courses, the Royal British Legion is still hearing complaints from former military that their qualifications are not recognised in the civil sector.

Clearly, any suggestion that many armed forces personnel cannot cope easily with civilian life, and may even figure prominently among the homeless,[12] must make it harder for the military to recruit the people they need. It also damages the reputation of the military among the wider public.

In 2017, in addition to enabling more flexible employment on a part-time, non-deployable basis in the military for experienced and qualified people, there was MoD consideration of an expansion in the number of military people moving into the civilian sector to gain valuable experience and/or to allow them a more stable period of employment when their family was at a particular stage.

This could leave open the possibility of a return to military service when their personal circumstances were easier. Should such a change take place, it would have the effect, albeit on a modest basis, of exposing the military to more of the wider community.

The military on operations

Public attitudes towards the military are particularly important when forces are deployed on hazardous operations where serious injury and even death are real possibilities for the people concerned. It is widely recognised that, for morale and operational effectiveness considerations, deployed forces need to feel that they have the support of their publics at home. Awareness that they are following government instructions is not enough.

Every operation is unique, with all that implies for the specific approach to building and sustaining support for it among the public and the media. But a generic issue for the defence sector is that, when organised armed forces clash with other organised armed forces, people invariably get injured and killed. In 1991, when the UK along with the United States and others was preparing to liberate Kuwait, the calculation was that three field hospitals would be needed to deal with casualties. Campaign planning for the military routinely addresses the replacement of casualties with fresh individuals.

This way of thinking does not apply in any other area of British life, including the other emergency services. Indeed, the law is strict on employers providing a safe place of work with every effort made to minimise the probability and impact of accidents.

In this context, and after large scale wars in 1991 and 2003 when very few soldiers died, the government's challenge has been to persuade the public that casualties are, to a degree, an inherent part of the military mission. There is no doubt that attitudes towards casualties have changed drastically in the era since 1945 and arguably even since 1982. Historical sources vary on the precise number of British military deaths during the Second World War but they averaged at least 147 a day.[13] The UK lost 255 personnel liberating the Falklands in a couple of months in 1982, 763 in the Northern Ireland conflict between 1969 and 2003, but 456 over the 12 years between 2003 and 2015 in Afghanistan. Every one of all these must have been a tragedy for the fallen's family and loved ones, but the casualty rate loomed largest in public consciousness in the case of Afghanistan.

A further feature of the Afghan conflict was the outstanding and constantly improving treatment of the wounded in theatre, which meant many people survived with injuries that would previously have been fatal. Yet the treatment of repatriated people in the UK, initially within the Selly Oak National Health Service in Birmingham, was widely seen as inadequate and unsuitable for returning soldiers.

In the eyes of many of the public, the soldiers killed and injured in the sustained campaigns in Iraq and Afghanistan seemed often to be regarded by the public as victims of government decisions rather than as soldiers who knowingly accepted the risks and rewards of the job.[14] Moreover, in the cases of both Iraq and Afghanistan, the British government could not point to any unambiguous metrics of 'success' or 'victory' as a consequence of the fighting and losses.

The expense and casualties of these operations brought relations between the military and society into sharp relief.[15]

The Military Covenant

None of the above issues were entirely novel and the MoD had long sought to manage them through a variety of public relations activities, including support for television documentaries about the military and recruitment activities that also served to tell the wider public about the military: the Royal Navy recruitment film of 2000, *A Cog in a Machine*[16] should be recognised as a classic: through images and a few words it presents the Navy as doing important work, focused on humanitarian action, offering important responsibility to all, being multi-racial and non-sexist, and offering a positive social dimension, all in less than 40 seconds. The Red Arrows have long delivered across the country and indeed the wider world an emphasis on the skills of the RAF. There were single service engagement teams that toured the country giving presentations about the activities of their branch of the armed forces,[17] and the MoD maintained a significant public relations office.

However, in 2011 the government decided to go further in terms of both its own commitments and the stimulation of more links between the military and the civilian world. The Military Covenant was placed on a legislated basis in 2011. Previously, the idea of a military covenant and that of the 'psychological contract' between the government and members of the armed forces had centred on the idea that, in return for the domestic disruption and risks of military life, the government would do all it could to assist injured soldiers and those leaving military service: 'you look after us so we will look after you'.

In the face of scepticism about the extent to which the government was living up to its side of the bargain, including exposure of the lack of appropriate care for wounded soldiers and the need for and growth of charities led by Help for Heroes, Prime Minister Cameron chose to incorporate government responsibilities and commitments into legislation in 2011, so that Covenant-based mechanisms became a key aspect of defence's relations with civil society.

It should be no surprise that a core paragraph of the Military Covenant reminds the public about the armed forces and their contribution.

> The first duty of government is defence of the realm. The Armed Forces fulfil that responsibility on behalf of the Government, sacrificing some civilian freedoms, facing danger and, sometimes, suffering serious injury or death as a result of their duty. Families also play a vital role in supporting the operational effectiveness of our Armed Forces. In return the whole nation has a moral obligation to the members of the Naval service, the Army and the Royal Air Force. They deserve our respect and support and fair treatment.

Thus, the Covenant was about reinforcing 'respect and support' for the Armed Forces and making sure they received 'fair treatment'.

The Covenant in itself involved a programme of change rather than immediate amendments to behaviour, alongside a commitment to publish an annual report on its implementation.

The envisaged changes involved areas for:

- Government action;
- Private sector action;
- Action at the local 'community' level.

Regarding the 'fair treatment' agenda, the government domain included issues where government had direct responsibility, such as medical care for the physically wounded and those psychologically scarred, and the legislative changes needed to ensure that soldiers were not disadvantaged by their time in the armed forces with regard to publicly-provided services and regulations.

At the private sector and the community levels, there was a twofold aim: ending the disadvantage of and building support for the military.

Ending the disadvantaging of the military

The first was to enlist the support needed outside central government to ensure that military personnel were not disadvantaged while in uniform and then afterwards in areas as diverse as house purchasing, mobile phone contracts, education, healthcare, and, of course, employment. This agenda was concerned with the provision of 'fair treatment'.

There was recognition that many rules and criteria, not least for access to health, education, and housing benefits had been set without consideration of the particular employment practices of the military. A soldier or a soldier's family member could be rising up a waiting list for an NHS treatment in one region, but then be transferred to the bottom of another region's list if that person was posted. Explorations of the Military Covenant identified a number of areas where public and private sector regulations and practices did have negative consequences for military people. For instance, a military person signing a three-year mobile phone contract and then being posted overseas was tied to the payment commitments despite not being able to use the phone. Military personnel posted to Germany or elsewhere whose mailing address was a BFPO postcode could subsequently struggle to obtain a mortgage because they could not provide a history of previous addresses: mortgage lenders did not recognise the BFPO system. A soldier who did buy a UK house for personal use but then had to move out because of an unexpected posting could not let the house to a tenant without re-negotiating the mortgage into a buy-to-let arrangement.

Some of these problems were not unique to the military, but a private-sector employer would probably finance any new arrangements needed for a mobile phone and house moves, and in any case private-sector employees usually have some degree of choice about where they work.

Building support for the military

The second was to generate increased emotional and practical commitment to support the armed forces and the work that they did, at the same time discouraging any view that the armed forces should be pitied. The government wanted the armed forces to have respect rather than pity or even sympathy.

Implementing the Covenant

To bring more people into contact with the military and to generate support at the popular level, the MoD has specified an Armed Forces Day involving locally-organised events around the country to celebrate the troops of the past, present, and future. Held in the summer, this is a celebratory event that is not to detract from the sombre reminders of sacrifice that is Armistice Day. The MoD has also directed the armed services to contribute to local events and celebrations whenever possible.

The government also set up a structure for non-legally binding but public commitments to be made by individual private organisations and by coalitions of relevant parties (which obviously included local governments) at the local 'community' level. The precise content of each 'Corporate Covenant' agreement and each 'Community Covenant' agreement, beyond a small core element, was a matter for the parties concerned.

In terms of Community Covenants, there was a rapid take-up and more or less all local governments quickly committed themselves in one form or another. The number of firms signing Corporate Covenants rose rapidly and, by the end of 2016, exceeded 1,500. In 2016 the government abandoned the distinction between Corporate and Community Covenants and merged them into a single group.

Many of both sorts of agreements, as well as generally expressing support for the armed forces, dealt with stances on the employment of former military personnel. Beyond the obvious commitment not to discriminate against retired military people, some others went further and offered advice on completing job applications. Some firms guaranteed that they would interview all ex-military candidates.

In terms of support for the uniformed force as a whole, of particular significance for the MoD was to win backing from employers for the reservists system and to minimise the chances that employers would be a barrier to people contemplating and practising reservist service. Many corporate Covenants address reservists, with a minority offering even paid leave for their employees to undertake days of reservist training.[18]

Through Covenant agreements, the government encouraged wider society to demonstrate its support and respect for the military through the granting of some privileges unavailable to others. The Defence Discount Service became the most obvious manifestation of such privilege.[19] In the United States, preferential treatment of soldiers can be demonstrated in conspicuous ways. Those using flights with American carriers may well have noticed that 'serving military' are specified as early boarders of the aircraft.

The Covenant initiative was about much more than improving the links between the military and the wider UK society. It dealt significantly with what in companies would have been seen as Human Resource Management issues, including for support and retention. It was also about giving the government's aspiration to attract more, and rely more on, reservists a better chance of success.

As the government-sponsored independent reports on the Military Covenant and other research indicates, the Covenant initiative made significant progress, but the MoD needed to be more pro-active in some areas. Our study found that these included better MoD guidance for companies that wished to contribute through service charities, of which there may be thousands. There was some evidence that some companies signed Covenants largely as a public relations gesture and to have another logo to put on their website and stationery. Specific commitment in those cases needed to be built and it was argued that the MoD should seriously consider expanding its Covenant office, which in 2017 was largely concerned with registering the agreements made.

On the issue of employment after military service, the Covenant appears reluctant to address the reality that the employment possibilities for ex-military personnel vary very much with their roles and related qualifications when in uniform. People of any rank with strong technical backgrounds in areas such as communications, electronics, engineering or in professional areas including accounting, logistics, and project management, predictably find it easier to find work than those who have served predominantly in the infantry where specialist skills focus on the use of armaments. The readiness of at least some civilian employers to take on senior ex-military in high-level roles is notable. In 2017 the Chief Executive of Transport for Greater Manchester is a former two-star officer who had gone initially from the RAF to running two London underground lines. After the MoD outsourced many of its logistics storage and distribution responsibilities to Leidos, a US firm, that firm then appointed another former RAF two-star to be its chief executive. A former Army colonel in a private business was running a major MoD accommodation project around Aldershot. While these represent merely anecdotal illustrations, they illustrate that leaving the military does not mean that being the bursar of a private school is the only employment option.

Major research in 2016 by the Royal British Legion reinforces these arguments, finding both that 85 per cent of UK service personnel (who used the MoD's Career Transition Partnership) were employed within six months of leaving the service, and that military veterans were twice as likely as the general population to be unemployed. A particular problem with infantry personnel who both join the service and leave without basic GCSE qualifications in English and Maths was noted. Some prejudice remains.[20]

More widely, the study found that employers often have inappropriate stereotypes and perceptions about military service and what it entails, signalling the need for the MoD to be pro-active with Covenant signatories in order to break down prejudice. The Covenant should not be viewed as pressing companies to employ veterans as a favour or reward for what they contributed in uniform, but as a means of building awareness of the qualities and positive experience of ex-military personnel.

While some stereotyping of military people is that they have limited self-control and rely on someone 'barking orders', opinion surveys consistently indicate the UK armed forces have a relationship with the national public that is extremely positive, with more than 80 per cent of the population having a high or very high opinion of the armed forces and more than 90 per cent expressing support for the people who served in Afghanistan and Iraq (despite less than 60 per cent endorsing the wisdom of the actual missions).[21] This appreciation of the armed forces is a solid basis for the further reinforcement of all the purposes of the Covenant which undoubtedly remains unfinished business.

While respect for the armed forces is high, this does not translate into clear support for increased defence spending, indeed the defence allocation 'sits roughly in the middle between those budgets that voters want to protect or cut the most'.[22] Additionally, and of significance for recruitment, only slightly more than 30 per cent of parents would be happy to see their child take up a military career.[23]

A key element must be the transition process between military and civilian life: defence cannot expect that the military enjoy status and appreciation for being 'special' because of unusual work that they do and the commitments that they make, and then hope that such people will be perceived as fitting smoothly into civilian occupations. Indeed, many really do not transition comfortably as the availability of former soldiers to work for private military and security firms attests. The Career Transition Partnership supported by the MoD therefore needs to move beyond supporting the acquisition of qualifications, articulation of transferable skills, the preparation of curricula vitae, and advice on behaviour at interviews. It needs to address both the accuracy of military stereotyping and persuasion that ex-military personnel can and do leave behind some elements of service expectations and behaviour when they take off their uniforms.

Conclusion: the implications of the 'Whole Force'

The experience of the protracted campaigns in the Balkans, Iraq, and Afghanistan was that the armed forces needed to deploy a range of people comprising full-time military, reservists, civil servants, and contractors in order to sustain the operation. The term 'Whole Force' was generated to summarise this arrangement and the end of the Afghan operation prompted the need to articulate, plan, and control it on a more systematic basis. The limited financial resources available for defence meant that it was too expensive to maintain a full-time comprehensive force that in peace-time would be focused only on training. There was thus a commitment in the 2015 SDSR to increase the size of the reserves to 35,000, a commitment that has proven difficult to carry through despite the contribution of the Military Covenant initiative discussed above.

However, one implication of the 'Whole Force' and the greater use of reserves is that more people could be linked to the military, if only on a part-time basis. On the other side of the coin, the armed forces are looking at releasing more of their full-time people into the civilian world to secure valued experience or perhaps just to enjoy a settled home life at a time when they may have pre-school or school-age children. Those getting valuable experience already include military doctors and other medical personnel, but the list could well widen to include cyber staff, aircraft technicians, and so on. Moreover, leaving the military may not be for good, as people may be happy to come back once their children have reached a certain age. The military problem is to maintain access to the skilled people in whose training it has invested, but a side effect could be to increase the number of individuals with military responsibilities and experience who are in day-to-day contact with wider society.

Notes

1 John Herz, *International Politics in the Atomic Age* (New York, Columbia University Press, 1959).
2 *The Economics of Adam Smith*, www.publicbookshelf.com/public_html/Outline_of_ Great_Books_Volume_I/adamsmith_bib.html, accessed 18 July 2017.
3 Adam Smith, *The Wealth of Nations*, Book V: On the Revenue of the Sovereign or Commonwealth, Chapter I: On the Expenses of the Sovereign or Commonwealth, Part I: On the Expense of Defence; www.marxists.org/reference/archive/smith-adam/ works/wealth-of-nations/book05/ch01.htm, accessed 3 April 2017.
4 Joseph S Nye Jnr, 'East Asian Security: the Case for Deep Engagement', *Foreign Affairs*, July-August 1995, www.foreignaffairs.com/articles/asia/1995-07-01/east-asian-security-case-deep-engagementp, accessed 18 July 2017.
5 Lindsey A Hines, Rachael Gribble, Simon Wessely, Christopher Dandeker, and Nicola T Fear, 'Are the Armed Forces Understood and Supported by the Public? A View from the United Kingdom', *Armed Forces & Society*, 41(4), 2014, pp. 1–26. www.kcl.ac.uk/kcmhr/publications/assetfiles/2014/Hines2014.pdf.
6 John Louth and Trevor Taylor, 'Beyond the Whole Force: The Concept of the Defence Extended Enterprise and its Implications for the Ministry of Defence', *RUSI Occasional Paper*, 2015.
7 https://en.wikipedia.org/wiki/List_of_military_veterans_in_British_politics;   Paul Goodman, 'Over 50 MPs have served in the armed forces …' *Daily Telegraph*, 15 November 2013; George Parker and Jim Pickard, 'A New Mission: Soldiers as MPs', *Financial Times Magazine*, 31 July 2015, www.ft.com/content/353a6ace-356d-11e5-b05b-b01debd57852?mhq5j=e1, all accessed 17 July 2017.
8 https://gcs.civilservice.gov.uk/about-us/, accessed 17 July 2017.
9 Ministry of Defence, '£135M infrastructure contract marks milestone in UK F35 programme', News story, 21 June 2017, www.gov.uk/government/news/135m-infrastructure-contract-marks-milestone-in-uk-f35-programme, accessed 22 June 2017.
10 See for instance, Christopher Hope, 'Philip Hammond blows £7.4million on botched plan to privatise defence procurement', *Daily Telegraph*, 10 December 2013, www. telegraph.co.uk/news/uknews/defence/10509188/Philip-Hammond-blows-7.4million-on-botched-plan-to-privatise-defence-procurement.html, accessed 14 February 2015;

Ben Farmer, 'MOD spends £66 million on consultants as Armed Forces cut', *Daily Telegraph*, 9 January 2014, www.telegraph.co.uk/news/uknews/defence/10559780/MOD-spends-66-million-on-consultants-as-Armed-Forces-cut.html, accessed 10 January 2015; Ian Drury, 'MoD spent millions of pounds on 6,000 pistols which it ditched after just five years', *Daily Mail*, 22 January 2014, www.dailymail.co.uk/news/article-2543694/MoD-spent-millions-pounds-6-000-pistols-ditched-just-five-years.html#ixzz4nB7t4XsN, accessed 4 February 2017; Guardian, 'Ministry of Defence "wasted millions on failed computer system"', *Guardian*, 14 January 2014, www.theguardian.com/uk-news/2014/jan/14/ministry-of-defence-failed-computer-system, accessed 18 July 2017.

11 See, for instance, Ministry of Defence, *Annual Report and Accounts 2015 to 2016* (London: The Stationery Office, 2016); Ministry of Defence, 'The Defence Equipment Plan 2016', January 2017, www.gov.uk/government/publications/ministry-of-defence-annual-report-and-accounts-2015-to-2016, accessed 18 July 2017; Ministry of Defence, 'Statistics at MoD', www.gov.uk/government/organisations/ministry-of-defence/about/statistics, accessed 18 July 2017.

12 Ben Glaze, '9,000 ex-service personnel homeless after leaving the military', *Mirror*, 21 July 2013, www.mirror.co.uk/news/uk-news/9000-ex-service-personnel-homeless-after-2071049, accessed 23 October 2016; Neal Keeling, 'Former soldier reveals plight as figures show huge rise in homeless ex-servicemen and women in Salford', *Manchester Evening News*, 26 June 2015, www.manchestereveningnews.co.uk/news/greater-manchester-news/former-soldier-reveals-plight-figures-9526327, accessed 20 April 2016; The Royal British Legion, 'UK Veterans and Homelessness', http://media.britishlegion.org.uk/Media/2283/litrev_ukvetshomelessness.pdf, accessed 11 January 2018; Jeremy Swain, 'The Bullshit Detector: Investigating a report into homelessness amongst former armed forces personnel', 30 July 2013, http://jeremyswain.blogspot.co.uk/2013/07/the-bullshit-detector-investigating.html, accessed 15 April 2015.

13 The war lasted just over 2,220 days and the lowest figure found for military casualties was 320,000 (http://secondworldwar.co.uk/index.php/fatalities). The highest was 382,000 British military killed (www.secondworldwarhistory.com/world-war-2-statistics.asp), all accessed 18 July 2017.

14 Rachael Gribble, Simon Wessely, Susan Klein, Christopher Dandeker and Nicola T Fear, 'The UK's armed forces: public support for the troops but not their missions?', in A Park, E Clery, J Curtice, M Phillips and D Utting, *British Social Attitudes: the 29th report* (London: NatCen Social Research).

15 Helen McCartney, 'The military covenant and the civil–military contract in Britain', *International Affairs*, March 2010, 86:2, pp. 411–28; Anthony Forster, 'Breaking the covenant: governance of the British army in the twenty-first century', *International Affairs*, November 2006, 82:6, pp. 1043–57; Timothy Edmunds, 'British civil–military relations and the problem of risk', *International Affairs*, March 2012, 88:2, pp. 265–82.

16 Available on YouTube at www.youtube.com/watch?v=yf3hPzMQfos, accessed 17 July 2017.

17 www.royalnavy.mod.uk/news-and-latest-activity/public-relations/rnpt; www.raf.mod.uk/presentationteam/; www.army.mod.uk/engagement/.

18 John Louth, Trevor Taylor and Lauren Twort, 'The Art of the Covenant: The Armed Forces Covenant and the Role of the Commercial Sector', *RUSI Occasional Paper*, 2016, https://rusi.org/publication/occasional-papers/art-covenant-armed-forces-covenant-and-role-commercial-sector, accessed 18 July 2017.

19 Defence Discount Service, www.defencediscountservice.co.uk/, accessed 1 January 2018.

20 www.britishlegion.org.uk/media/5035/deployment-to-employment.pdf, accessed 17 July 2017.

21 www.bsa.natcen.ac.uk/media/1150/bsa29_armed_forces.pdf; see also Joel Faulkner Rogers, 'Report on British Attitudes to Defence, Security and the Armed Forces', 25 October 2014, https://yougov.co.uk/news/2014/10/25/report-british-attitudes-defence-security-and-arme/, accessed 3 February 2016.
22 Joel Faulkner Rogers, 'Report on British Attitudes to Defence, Security and the Armed Forces', 25 October 2014, https://yougov.co.uk/news/2014/10/25/report-british-attitudes-defence-security-and-arme/, accessed 3 February 2016.
23 Ibid.

Bibliography

Defence Discount Service, www.defencediscountservice.co.uk/, accessed 1 January 2018.

Drury, I (2014), 'MoD spent millions of pounds on 6,000 pistols which it ditched after just five years', *Daily Mail*, 22 January 2014, www.dailymail.co.uk/news/article-2543694/MoD-spent-millions-pounds-6-000-pistols-ditched-just-five-years.html#ixzz 4nB7t4XsN, accessed 4 February 2017.

Edmunds, T (2012), 'British civil–military relations and the problem of risk', *International Affairs*, March 2012, 88:2, pp. 265–82.

Farmer, B (2014), 'MOD spends £66 million on consultants as Armed Forces cut', *Daily Telegraph*, 9 January 2014, www.telegraph.co.uk/news/uknews/defence/10559780/MOD-spends-66-million-on-consultants-as-Armed-Forces-cut.html, accessed 10 January 2015.

Forster, A (2006), 'Breaking the covenant: governance of the British army in the twenty-first century', *International Affairs*, November 2006, 82:6, pp. 1043–57.

Glaze, B (2013), '9,000 ex-service personnel homeless after leaving the military', *Mirror*, 21 July 2013, www.mirror.co.uk/news/uk-news/9000-ex-service-personnel-homeless-after-2071049, accessed 23 October 2016.

Goodman, P (2013), 'Over 50 MPs have served in the armed forces …' *Daily Telegraph*, 15 November 2013, www.conservativehome.com/parliament/2013/11/over-50-mps-have-served-in-the-armed-forces-here-are-more-than-30-of-them-together.html.

Gribble, R, Wessely, S, Klein, S, Dandeker, C, and Fear (2012), 'The UK's armed forces: public support for the troops but not their missions?', in A Park, E Clery, J Curtice, M Phillips, and D Utting, *British Social Attitudes: the 29th report* (London: NatCen Social Research).

Guardian (2014), 'Ministry of Defence "wasted millions on failed computer system"', *Guardian*, 14 January 2014, www.theguardian.com/uk-news/2014/jan/14/ministry-of-defence-failed-computer-system, accessed 18 July 2017.

Herz, J (1959), *International Politics in the Atomic Age* (New York, Columbia University Press).

Hines, LA, Gribble, R, Wessely, S, Dandeker, C, and Fear, NT (2014), 'Are the Armed Forces Understood and Supported by the Public? A View from the United Kingdom', *Armed Forces & Society*, 41(4), pp. 1–26.

Hollander, S (1973), *The Economics of Adam Smith* (University of Toronto Press), www.publicbookshelf.com/public_html/Outline_of_Great_Books_Volume_I/adamsmith_bib.html, accessed 18 July 2017.

Hope, C (2013), 'Philip Hammond blows £7.4million on botched plan to privatise defence procurement', *Daily Telegraph*, 10 December 2013, www.telegraph.co.uk/news/uknews/defence/10509188/Philip-Hammond-blows-7.4million-on-botched-plan-to-privatise-defence-procurement.html, accessed 14 February 2015.

Keeling, N (2015), 'Former soldier reveals plight as figures show huge rise in homeless ex-servicemen and women in Salford', *Manchester Evening News*, 26 June 2015, www.manchestereveningnews.co.uk/news/greater-manchester-news/former-soldier-reveals-plight-figures-9526327, accessed 20 April 2016.

Louth, J and Taylor, T (2015), 'Beyond the Whole Force: The Concept of the Defence Extended Enterprise and its Implications for the Ministry of Defence', *RUSI Occasional Paper*.

Louth, J, Taylor, T and Twort, L (2016), 'The Art of the Covenant: The Armed Forces Covenant and the Role of the Commercial Sector', *RUSI Occasional Paper*, 2016. https://rusi.org/publication/occasional-papers/art-covenant-armed-forces-covenant-and-role-commercial-sector, accessed 18 July 2017.

McCartney, H (2010), 'The military covenant and the civil–military contract in Britain', *International Affairs*, March 2010, 86:2, pp. 411–28.

Ministry of Defence (2016), *Annual Report and Accounts 2015 to 2016* (London: The Stationery Office).

Ministry of Defence (2017), '£135M infrastructure contract marks milestone in UK F35 programme', News story, 21 June 2017, www.gov.uk/government/news/135m-infrastructure-contract-marks-milestone-in-uk-f35-programme, accessed 22 June 2017.

Ministry of Defence (2017), 'The Defence Equipment Plan 2016', January 2017, www.gov.uk/government/publications/ministry-of-defence-annual-report-and-accounts-2015-to-2016.

Ministry of Defence, 'Statistics at MoD', www.gov.uk/government/organisations/ministry-of-defence/about/statistics, accessed 18 July 2017.

Nye Jnr, JS (1995), 'East Asian Security: the Case for Deep Engagement', *Foreign Affairs*, July–August 1995, www.foreignaffairs.com/articles/asia/1995-07-01/east-asian-security-case-deep-engagementp, accessed 18 July 2017.

Parker, G and Pickard, J (2015), 'A New Mission: Soldiers as MPs', *Financial Times Magazine*, 31 July 2015, www.ft.com/content/353a6ace-356d-11e5-b05b-b01debd578 52?mhq5j=e1, all accessed 17 July 2017.

Rogers, JF (2014), 'Report on British Attitudes to Defence, Security and the Armed Forces', 25 October 2014, https://yougov.co.uk/news/2014/10/25/report-british-attitudes-defence-security-and-arme/, accessed 3 February 2016.

Smith, A (1776), *The Wealth of Nations*, Book V: On the Revenue of the Sovereign or Commonwealth (London, Seedbox Press, 2011 edition).

Swain, J (2013), 'The Bullshit Detector: Investigating a report into homelessness amongst former armed forces personnel', 30 July 2013, http://jeremyswain.blogspot.co.uk/2013/07/the-bullshit-detector-investigating.html, accessed 15 April 2015.

The Royal British Legion, 'UK Veterans and Homelessness', http://media.britishlegion.org.uk/Media/2283/litrev_ukvetshomelessness.pdf, accessed 11 January 2018.

www.royalnavy.mod.uk/news-and-latest-activity/public-relations/rnpt.

www.raf.mod.uk/presentationteam/; www.army.mod.uk/engagement.

www.britishlegion.org.uk/media/5035/deployment-to-employment.pdf, accessed 17 July 2017.

www.bsa.natcen.ac.uk/media/1150/bsa29_armed_forces.pdf.

https://gcs.civilservice.gov.uk/about-us/, accessed 17 July 2017.

9   Defence as teamwork and partnering

Introduction

This chapter introduces the idea of defence as a function generated by multiple organisational partners drawn from the public and private sectors. In many ways this seems counterintuitive. When the casual reader considers ideas around defence or the armed forces, she or he tends to focus upon trained people in military uniform within regiments, on ships, or, perhaps, crewing or maintaining aircraft. At this first glance, defence seems to be the preserve of the Royal Navy, British Army, and Royal Air Force. What is the private sector contributing to this force mix?

Well, the answer of course is a great deal. When people in the UK initially join the military for a career they are drawn, initially, from broader society – perhaps from school, university, or a previous civilian job. Consequently, some educational and behavioural skills and values are forged in society prior to military service. Second, once a woman or man has completed their time in one of the Services, she or he typically will have a regular reserve obligation whereby they can be recalled to the regular forces. What impact would this have on a future employer? Third, a number of civilians volunteer as reservists to train at evenings or weekends for deployment with the regular military. They need the tacit agreement of their normal employer to undertake these activities along with, presumably, the support of their families. Last, much of the technologies, equipment, consumables, and even ideas that are utilised by the armed forces are formed and housed in businesses as suppliers to the Ministry of Defence. These relationships can be a complex symbiotic web of commercial obligations and values or emotions around notions of nationhood, loyalty, and service. The generation of defence capability is a very complicated public–private relationship indeed. As Sir Peter Luff states: 'Without industry, or even broader society, there can be no armed forces so there would be no national defence. Partnering between the private and public sector is everything.'[1]

Chapter objectives

By the end of this chapter the reader will understand:

1   How the subject of UK defence has evolved from essentially a Cold War posture to more complex operations of choice and the impact this has had on private sector suppliers;
2   The arguments around the idea of defence as a public–private partnered imperative;
3   The notion of the UK 'Whole Force'.

Chapter structure

This chapter starts with a short historical perspective on the relationship between the military and, principally, industry in the context of the Cold War and a bipolar world between the countries of NATO and the Warsaw Pact. It goes on to consider just what we mean by partnering within defence and to draw upon some examples to illustrate the argument. Thereafter, the chapter introduces the UK government's concept of the 'Whole Force' which, supposedly, acts as a conduit for the realisation of the British modern ambition for defence partnering.

Towards partnering

The Cold War period from just after the end of the Second World War through to the collapse of European state communism in the late 1980s and early 1990s represented a quite remarkable break in modern history. Conventionally, a central tenant of the projection of military power was viewed as the ability to access during armed conflict a capable, secure, agile, and reliable source of supply of equipment, materials, and services necessary to sustain war. Britain's economy from the late 1930s to about 1945 was principally concerned with delivering this critical *national* and *military* imperative – the two were interchangeable.

In contrast, the central military task of the British associated with the Cold War was that of deterrence. The main national task was economic prosperity or reversing economic decline. Throughout this period the UK maintained a determinedly defensive posture to deter Soviet aggression. 'Europe was the most militarised region on the planet and the UK prepared a defensive strategy to withstand, collectively, a major onslaught from the Warsaw Pact towards the British homeland.'[2] Any action that the UK undertook during the Cold War period (operations in what was Malaya, Kenya, Cyprus, Borneo, Oman, and the Falklands) was an addendum to the core task of using escalatory military means to deter Warsaw Pact aggression in Europe and to protect liberal democracy.

So, prior to 1945, the UK was involved in a global conflict that was, in essence, a war of attrition between advanced global economies. The alliance that could produce, sustain, or replace the largest number and most effective equipment and materials for military usage would win. From the Cold War onwards, there was little need in the West to mobilise its industries for a protracted conflict of survival. Rather, deterrence throughout the Cold War was predicated on a state's ability to utilise weapons and materials that had already been ordered, manufactured, proven, and deployed. Glibly, for some, the imperative became stock management.

The concept of defence partnering

In defence management, an often-heard evocation is for government and industry to, somehow, become 'partners' in the generation of defence capabilities. For instance, in 2012 there have been a number of stories in the UK press reporting that defence procurement is about to be turned into a 'strategic partnership' through the appointment of a private-sector consultancy or service provider specialising in programme delivery to share with the MoD the management of Defence Equipment and Support (DE&S).[3] Indeed, Bernard Gray, the Chief of Defence Materiel, argued in November 2011 that the provision of such a strategic partner was a principal option he was exploring for Departmental reform.[4]

Moreover, it is reported that industry is to have a significant role to play in running Defence Business Services for the Department, an organisation that delivers to the MoD back-office support, such as finance and human resources. In addition, companies were asked on 4 January 2012 to register expressions of interest to manage, under a partnering initiative with the MoD, the sprawling defence estate portfolio that is the Defence Infrastructure Organisation.[5] Understanding, therefore, what 'partnership' or 'partnering' actually mean to both the MoD and industry is critical if analyst and citizen alike are to make sense of what seems a major aspect of defence management and ongoing public-sector reform.

Consequently, this chapter was stimulated by three interlocking considerations. First, from the change management agenda of the Labour government's Smart Procurement Initiative onwards, beginning in 1998, there has been a clear policy preference from within the MoD for the Department to deliver certain defence capabilities or activities, somehow, in partnership with industry. How this was to be done, or indeed how this has been achieved between then and now, is certainly worthy of a short review. Second, under the Coalition government, the MoD launched in late 2010 a procurement programme known as the Chemical, Biological, Radiological and Nuclear (CBRN) Key Strategic Partnership for the business transformation of this nationally significant sector. This was a bounded and costed year-long assessment phase programme between the MoD and a number of defence companies to generate, inter alia, the business case for a long-term partnered solution between government and industry to provide national Chemical, Biological and Radiological capabilities. This programme was hailed as a significant test-case for future strategic partnering initiatives between the MoD and industry. As it concluded at the end of 2011, unpicking the results of this programme, or any lessons learnt from it, should provide a timely insight into the processes and purposes of partnering pan-defence.

Third, and as stated above, there is the widespread media coverage of the Chief of Defence Materiel's options for changes to DE&S, whereby the MoD and industry are expected to partner with each other in the delivery of defence procurement for the UK.[6] Given the large sums of money concerned,[7] and the significance of defence acquisition to the success or otherwise of the military component, understanding the 'partnering' phenomenon and its place within defence policy is now critical.

This chapter makes a simple argument. Defence policy makers and corporate executives need to be clear-eyed about programme intent and required benefits when they use terms such as 'partnering' or 'partnership'. Clarity around the operationalisation of that intent is very important, through costed, bounded, and properly resourced delivery programmes. In short, there is a world of difference between, on the one hand, a glib, uncritical partnering posture and, on the other, the complexities and challenges of real delivery programmes to actually bring forward the required effects and benefits necessary for a nation's security. When the two are confused, the UK runs the risk of capability gaps and defence incoherence.

The work proceeds as follows: a treatment of the words 'partnership' and 'partnering' is initially offered, with the sense that they are fairly loosely defined and often interchangeable concepts. Thereafter, the MoD's preference for partnering is explored through programme examples such as the MBDA-led Team Complex Weapons (TCW) and QinetiQ's Long Term Partnering Agreement (LTPA) with the MoD. Third, the CBRN Key Strategic Partnership assessment phase programme is reviewed in order to glean lessons that can be applied from this self-styled partnering test-case. The chapter then concludes with an assessment of the key insights drawn from this review, and offers some recommendations.

A word of caution is required. This chapter does not seek to provide a comprehensive review of partnerships or partnering. Neither does it look to explore all defence activities. Instead, it is deliberately selective in its attempt to shape the debate within the boundaries of the projects studied. Through this focused treatment, lessons derived can be applied to the broader defence policy options that appear to be on the table today.

Partnership and partnering

There are two terms being used; that of 'partnership' and that of 'partnering.' The *Concise Oxford Dictionary* describes a partnership as 'a state of being a partner, where partner is a person who shares or takes part with another, especially in business, with shared risks and profits'.[8] To partner, therefore, is to join with another to deliver a considered set of outputs or outcomes. For example, *The Gower Handbook of Management* notes that, instead of conducting a particular activity through the medium of a corporation possessed of a specific legal identity, or a government department or other public body, a partnership is merely a form of organisation that binds two or more parties together in pursuit of a common purpose. More often than not, no separate legal entity is formed beyond the identity of the individual partners themselves, with some form of negotiation between the parties still necessary to articulate their respective responsibilities to each other.[9]

'Partnering' is a much looser term, and one derived from the Japanese model of post-war manufacturing especially within the automotive industry.[10] Toyota's policy, for example, rests upon three principal strands. First, the assembling organisation controls the relationship as senior partner (in Japan, the word 'parent' is used, with component manufacturers seen as 'children'), but specialist suppliers are recognised as being absolutely critical to the overall quality of the product. There is recognition of expertise and quality running throughout the supply chain that binds customer to prime contractor and, thereafter, to niche supplier, so that each is dependent upon the other, committed to this sense of quality, shared service goals, and common expectations. Second, and critically, the specialist abilities of the sub-contractors are explicitly valued as crucial to the success of the overall product. Components are not sourced by Toyota, rather partners from the supply chain are found, developed, and valued for the long-term.[11]

Thus, for some in the MoD, partnering is about a long-term non-adversarial affiliation between the Department and a particular company or companies within a specific sector of defence, whereby the relationship becomes central to the delivery of effective and affordable capabilities.[12] Why does the government seek such a relationship with industry to deliver perceived key defence services or effects? A conventional response seems to be as follows.[13]

First, it can be said that government and industry come together to, somehow, lever-in to public services the private money that government neither has itself nor can afford to generate through taxation or the money markets.[14] Second, partnered arrangements are perceived to generate private sector capacity and competency to potentially supplant or enhance public sector provision.[15] Third, at an operational level, these relationships are believed by some to deliver greater value for money than sole public-sector provision.[16] This is said to be achieved by government transferring to the private sector costs and risks that would otherwise be borne solely by the public.[17] Moreover, it is often assumed that industry is possessed of greater expertise, innovation, and efficiency than its potential partner from the public sector, and can manage these costs and risks much more effectively.[18]

These points, together, represent the conventional, rational explanation for the practice of governments forming some sort of partnering arrangement with companies from the private sector to deliver defined goods and services. Some academics suggest that there are, perhaps, four distinct types of arrangements:

- The Collaborative: the policy of equal resources being provided by partners in pursuit of shared goals;
- The Operational: some form of identified work-share within a specified programme or project, but with individual partners keeping their autonomy for decision-making and action;

- The Contributory: the sharing of financing for a particular activity, but not the operational control, and;
- The Consultative: advice sought from one partner to assist another in the delivery of a specified product or function.[19]

Whatever the particular arrangement, the critical components are clarity and a corresponding sense of trust between the partners. We shall come on to discuss shortly how characteristics like these contribute to the MoD's supposed quest for a 'partnered solution' within defence procurement. However, for now, it is important simply to note that there seems to be some kind of body of knowledge relating to notions of partnering and partnership that can be deployed by policy makers and officials as they seek to put in place effective, efficient and economic arrangements for engaging with industry to deliver defence capabilities. It is also fair to say that there are no clear, unambiguous, and commonly used definitions of 'partnering' or 'partnership' within defence, as the terms themselves seem always to invite further explanation. Instead, and intriguingly, the notion of partnering, though not explicitly defined, is seen by a number of officials in Whitehall and Abbeywood as an increasingly important objective in its own right, colouring, for example, the policy debate on the future of the DE&S organisation.[20]

The MoD and partnering

The modern narrative constructed around partnering between the MoD and industry is rooted in the Smart Procurement Initiative reforms launched as part of the Labour Government's Strategic Defence Review in 1998.[21] The MoD's Smart Acquisition Handbook, in 2002, described partnering as the development of a new, much more co-operative long-term relationship between the MoD and industry.[22] It was intended to offer a refreshing change from conventional MoD contracting mechanisms in that effective communications between the MoD project team and industry would, through partnering, build trust and minimise project risks and uncertainty.

Consequently, MoD's high-level guidance for the relationship with industry stated that:[23]

> We [MoD] have recognised that a partnering approach – building reliable links with industry – is often the best way of achieving the required performance.... The partner is given opportunities to achieve innovation and value for money. If he is successful and demonstrates he can meet the Department's needs efficiently, and cost effectively, the business of the partnership can expand.

One such commercial relationship is the Long Term Partnering Agreement (LTPA) between MoD and QinetiQ. It provides an 'innovative partnering approach to managing MoD's strategic test and evaluation capabilities' as well as some elements of defence training.[24] Under the contract, QinetiQ offers within specific performance parameters real-time and simulated test and evaluation services to the MoD for its platforms, systems, weapons, and components. The objective is to provide the Department with an accurate assessment of capabilities throughout their respective life-cycles, thereby increasing reliability and fitness for purpose in the mid to long-term.[25]

The LTPA is certainly a long contract, running for 25 years to 2028, with an MoD option to extend until 2053. It is specifically designed to be: 'An innovative partnering arrangement which will enable long-term planning to meet MoD's current, evolving and future T&E needs.'[26] The words 'innovative' and 'partnering' feature prominently in how the agreement is explained. Indeed, the governing partnering principles within the contract are said to be: teamwork, safety, investment, performance, and risk management. These terms are actually hard-edged commercial requirements and practices, detailing obligations between the MoD customer and the industry supplier. It is difficult to make out what the 'partnering' label adds to this commercial arrangement.

This is the basic challenge for those who champion the idea of partnering between the MoD and defence industry. The principles framing the relationship between the Department and QinetiQ within the LTPA are sound and sensible, and a number of individuals from both the Department and industry have spoken of the contract's success.[27] But it is built on strictly commercial lines with clearly defined requirements, performance obligations, and financial rewards, making it a challenge to understand what partnering in defence actually represents in the context of a commercial relationship and contract. Indeed, a retired, former senior executive from QinetiQ suggested that the onus on partnering within the LTPA was a branding and marketing issue only. The notion of partnering adds neither substantive structural, procedural, nor, indeed, even behavioural dimensions to the contract, but is instead a useful label to articulate an aspiration (albeit an important one) for good relations between the contractual parties.

Good relations, confidence, and trust within a contract are of course excellent qualities and highly prized within the commercial world.[28] If the MoD and defence companies choose to use the term 'partnering' as a handy code to signal a valuable commercial relationship, it is certainly conceivable that this could prove helpful to a broad variety of stakeholders. This proposition veers towards the problematic when this handy code becomes a championed organisational and operational solution in its own right; in other words, when 'partnering' becomes operationalised in the minds of decision makers to, somehow, offer a solution to complex organisational and structural challenges such as the future of DE&S.

Under the previous government, when the MoD's Defence Industrial Strategy (DIS)[29] was published in December 2005, the MoD and UK complex weapons industry were encouraged to develop a fresh approach to the development, acquisition, and support of complex weapons within the UK. Declining historical research and development investment in these sorts of weapons packages, plus the government's desire to retain appropriate operational sovereignty through the maintenance of matching sovereign industrial competencies and capabilities, demanded under the strategy a sector-wide approach. Indeed, at the time, much commentary and analysis focused on a 'pan-sector partnered solution'.[30]

In this context, though, the concept of partnering was significantly more than just a branding or marketing exercise. Team Complex Weapons (TCW) was formed by industry in 2006 in response to the sector's strategic challenges articulated by the DIS, continues to work effectively under the joint leadership of the MoD and MBDA, and has had Thales Air Defence, Thales Missile Electronics, Roxel, and QinetiQ as key members of the team. Good relations, confidence and trust aside, partnering for TCW is borne from robust contractual and organisational management principles. First, TCW runs through a complex, though highly visible, portfolio management agreement underscored by excellent scheduling, management accounting, and risk management techniques. Responsibilities and obligations are clearly understood, with companies and individuals alike held to account against a master schedule of work. Joint and regular reviews between the MoD and MBDA, and the wider team members, have led to openness and transparency at the portfolio level, but also within individual corporate budgets, estimating practices and pricing schedules. So that TCW, for many, provides an important example of how MoD and different companies from within a particular defence sector can work together sensibly, effectively, and in a partnered way. The key point is that any sense of partnering within TCW is not regarded as an objective or solution in its own right, but is just one element of a programme solution driven by an overt national requirement, profound technical and operational challenges, resource constraints, and, at one time, a limited and fragile knowledge base.

If the drafters of the DIS in 2005 foresaw TCW as providing the organisational vehicle for securing, protecting, and enhancing the UK complex weapons sector, they could scarcely be disappointed. The development of a £4 billion portfolio of complex weapons projects is already within the TCW pipeline, the military effort in Afghanistan has been proactively and successfully supported, and TCW's contribution to the Libyan campaign in 2011, whilst still hazy, seems to have been profoundly significant. Indeed, one UK commander, in a background briefing to RUSI, described TCW's direct support to operations in Libya as absolutely central to the military effort.[31]

Notions of partnering can be significant, therefore, when aligned with, and subordinate to, robust portfolio and project management skills, design and engineering competencies, and unambiguous commercial and contractual responsibilities. Not as an end in itself of course, but rather as a commitment from colleagues within MoD and industry to work together, openly, on complex, nationally significant problems.

The CBRN key strategic partnership

From this short review, it seems clear that there exists across the defence community in the UK both some kind of narrative of partnering between MoD and industry and the experiences of specific partnering and teaming initiatives such as the LTPA and TCW. It was inevitable, perhaps, for the MoD to come to the conclusion that it wished to explore whether the partnering idea in the abstract, conceptual and alone, could be the answer to a number of loosely defined ills and organisational challenges. Consequently, in 2010, an assessment phase to test the idea of a Key Strategic Partnership was launched – that of a major partnering programme with industry to manage and sustain CBRN force protection. This phase concluded at the end of 2011, the lessons from which are still being gathered and explored. It makes sense, though, to look at this concept given that it was often presented as a test-case for broader DE&S reform.[32]

The Key Strategic Partnership for CBRN sector transformation was, in many ways, a fairly straight-forward proposition.[33] The Department's CBRN Protection Delivery Team from within DE&S contracted with Serco Selex (Inform), KBR, and QinetiQ to form with MoD a joint team to undertake an assessment phase programme throughout 2011. This year-long programme was aimed at testing and evaluating the benefits, opportunities, and risks of the Key Strategic Partnership concept within the CBRN sector.

An initial question, inevitably, is just what was this concept? This has never been satisfactorily explained. Rather the Key Strategic Partnership *concept* became the *activities* within the assessment phase and the *activities* of the assessment phase became the *concept*.

Nonetheless, much work was organised around the main themes of:

- The business case – oversight and assurance;
- Benefits – value and realisation;
- Stakeholder management and communication;
- Through-life capability planning and delivery;
- Supplier network management, and;
- Joint team organisational solution, post-assessment phase.[34]

The objective was to submit a business case to the MoD Board at the end of the assessment phase in December 2011 outlining how the joint team of MoD and industry Key Strategic Partners could deliver CBRN capabilities in the long-term more efficiently and effectively than a public-sector comparator.

Throughout 2011 a significant range of activities were witnessed, including extensive industry days, new governance group meetings, and briefings to senior officials as the team members, in good faith and with considerable dedication, sought to deliver, somehow, this partnership. A question, though, seemed frozen on many colleagues' lips: for what purpose?

When this was first asked at an industry day in Birmingham on 19 January 2011, the response from MoD officials was that partnering and partnership was one and the same thing, and that partnering between the MoD and industry was a credible outcome in its own right.[35] It was one that, by definition, would deliver economies and efficiencies to the UK. However, the MoD's own unclassified Concept of Analysis[36] for this work described the assessment phase as a

'Category A' project to offer capability management within a rebalanced DE&S. It is not an exaggeration to suggest that there seems to be at best some confusion around what the planning intent for this programme actually was. Moreover, there was ambiguity around the success criteria for the assessment phase. For the industry partners it was the delivery of the business case which would release the funds from MoD for a long-term partnered solution for UK CBRN capabilities. For some in the Department, however, the focus was on industry advice to MoD and some kind of specialist manpower substitution for skills gaps in Abbeywood. That this ambiguity continued throughout the programme is, perhaps, to no one's credit, questioning the ability of the MoD and industry to work well together in this sector.

Given this, and not surprisingly, no main gate business case submission for the Key Strategic Partnership concept was either offered or passed in December 2011. Rather, colleagues reported that a fatal lack of clarity over programme requirements, methodology, project management practices, and matching resources left this partnering concept floundering.[37] As a test-case for wider DE&S reform it hardly looks promising.

MoD and industry partnering: flirtation or marriage?

Yet some significant, important, and timely lessons can be learnt, both from the experiences of the CBRN Key Strategic Partnership initiative and this broader short review. First, there is an established academic and practitioner body of knowledge relating to ideas of partnering and partnership that seems to emphasise the criticality of strong portfolio and project management techniques and competencies. The MoD and industry can come together to deliver effective solutions for defence that are collaborative in nature, where equal resources are committed by partners in pursuit of shared, clearly identified goals, and where there is an obvious form of work-share articulated across a broad portfolio of interrelated projects. Individual partners can keep their respective identities and even their autonomy and yet still work in an open and frank way, utilising common tools and techniques, where partners are consultative, considerate, and comfortable working within the collective. Some analysts and practitioners would argue that this sort of partnering arrangement already exists as a proven model within TCW, a model that is purposeful, unambiguous, yet overtly commercial.

So, partnering as a concept seems best when it is enwrapped by robust programme management techniques. It becomes less meaningful when deployed merely as a branding or marketing tool, and is devoid of all relevance when partnering becomes an objective in its own right. Consequently, DE&S, just like any other function of government, cannot be effectively reformed simply by decision makers playing the partnering card. An organisational objective, sense of critical outcomes, and set of requirements are necessary, linked to long-term sources of finances, clear roles and responsibilities, and robust reward mechanisms where risks are shared and opportunities mutually pursued. Of course, the positive behaviours associated with the partnering label must have a part to play, but effective functional delivery is generated by clear-headed and robust portfolio and corporate management skills, not managerial slogans. It is to be hoped that, when decision makers and senior officials speak of defence reform, they have not been seduced by the pert phrase and flotsam of managerial fashion. The pretty, teasing words of a certain type of partnering can make for a pleasant, short-term flirtation, but they are not the stuff of a successful long-term marriage. That requires altogether stronger foundations.

The UK's 'Whole Force' approach

Since about 2010, the 'Whole Force' is the signifier applied to the UK's changing defence posture in which defence forces morph from being solely composed of a volunteer, professional army, navy, and air force to a partnered arrangement of regular military, regular reserves, volunteer reserves, sponsored reserves, private-sector contractors, and members of the civil service under varying contractual arrangements. As this author has written previously,[38] the common-sense view traditionally was that military operations were undertaken and supported by military personnel in the UK involving people from the Royal Navy, the British Army, and/or the Royal Air Force. Wars are fought by men and women in uniform not in factory overalls, proudly displaying the flag of their nation rather than the logo of their corporation.

Of course, in reality, businesses in the private sector contribute significantly to the generation of military capabilities through the development and manufacture of military equipment, its servicing, and the provision of training for military personnel in the use of materials and machines provided by the private sector. Also, a company can undertake multiple functions within a theatre of operations under contract to the military. Indeed, the final MoD report on operations in Iraq, during which about 1,500 civilian contractors were deployed to the Gulf region from 2003 until the end of the decade, states:[39]

> The very considerable success in delivering equipment against very demanding time and performance criteria owed much to the excellent contribution of contractors in the face of relatively late changes to the force composition and constraints on early consultation with industry.

In 2010 about 7,000 contractor employees were deployed by the UK on Operations *Telic* (the Gulf), *Herrick* (Afghanistan), *Calash* (Indian Ocean), and *Oculus* (the Balkans). Contractor Support to Operations (CSO) accounted at the same time for about 45 per cent of the UK overseas operational effort. Indeed, estimated annual CSO expenditures for 2010 came to around £2.6 billion.[40] The concept of the 'Whole Force' could be merely a manifestation of a trend towards the public–private partnering of defence on operations that characterises the UK's wars of choice in Iraq and Afghanistan in this century.

The strategic context for the development of the 'Whole Force' was a reduction in real terms in the defence budget of about 7.5 per cent over four years from 2010.[41] Indeed, in 2011, the Deputy Chief of the Defence Staff (Personnel and Training) captured emerging thinking around the 'Whole Force' Concept, which was portrayed as a comprehensive approach between uniformed and private sector partners to the delivery of required effects on the frontline. In future, this would be achieved by providing a blend of civilian, MoD civil service, and contractor personnel, deployed in harmony with sponsored reserves, the regular reserve, the volunteer reserve, and the regular armed forces. The trade-off for this optimum force mix was between readiness levels and the duration of operations.[42] The government's pictorial representation of the 'Whole Force' Concept, as at 2011, is given in Figure 9.1.

The claxon of the 'Whole Force' has been sounding, sometimes quietly, often more overtly, since 2010. Yet, to date, few substantial contracts relating to sponsored reserves, for example, and their role on the frontline appear to be in place between the MoD and defence contractors. The number and identification of posts in industry that should be sponsored reserves under the 'Whole Force' model remains to be quantified. As one senior industrialist described it, detailed and centralised planning for the 'operationalisation' of the 'Whole Force' by government remains a work in progress.[43]

Also, a lack of centralised planning and record keeping could lead to some interesting challenges within the 'Whole Force'. Let us assume that there is an individual employed by a defence contractor who has specific skills in nuclear propulsion design. These skills are now especially valuable given the

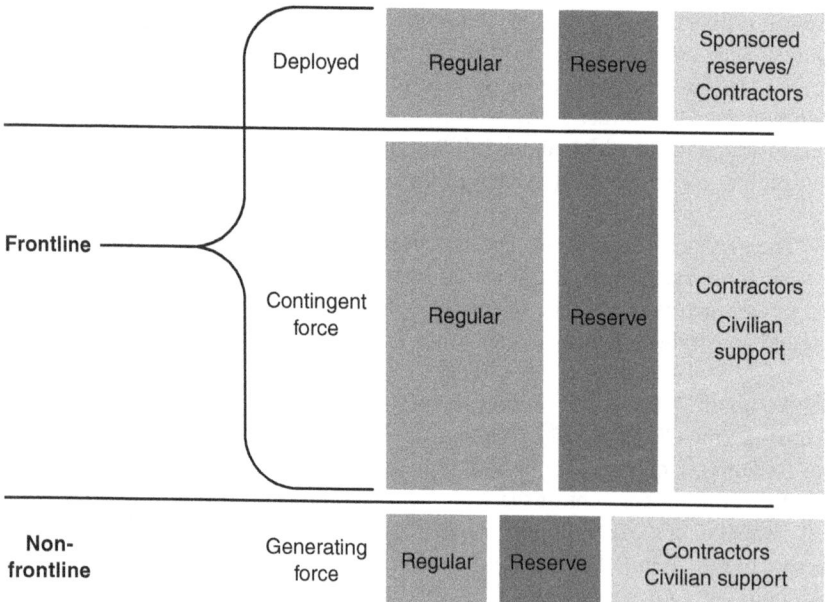

Figure 9.1 The 'Whole Force' concept mix.
Source: the authors (2018).

Dreadnought programme for the renewal of the UK's nuclear deterrence. The prime contractor responsible for this programme wishes to employ this skilled person and agrees a contract with his company for his services. This person, given his rare skills, currently sits in a sponsored reserve identified post to provide short-notice operational support to the Royal Navy. To complicate matters further, he is also a volunteer reservist in his spare time and is trained for service with the infantry. There are, in theory, three distinct and separate defence requirements placed on this one individual. Which role has priority and who decides? It seems pertinent to point out that no organisational structure and decision-making criteria appear to be in place for addressing the sort of scenario suggested above.

It is this notion of the 'Whole Force', developed during the early years of the Conservative-Liberal Democrat coalition government, that appears to have the same issues of detail and intent as identified with Joint Force 2025. This may be because the complexities of enterprise management, as the next section describes, are not embraced within UK defence planning and decision-making circles.

Conclusion

This chapter has explored a short recent historical perspective on the relationship between the British military and the civil, principally commercial industrial, sector. It has framed an understanding of defence within a public–private partnership and captured how the phenomenon has morphed into the concept of the 'Whole Force'. As we move on to explore British defence through the lens of the armed forces themselves we will see just how significant the notion of the 'Whole Force' was to become in the minds of defence planners and commanders.

For now, perhaps there are two thoughts that encapsulate the centrality of partnering to the practice of defence in Britain. As Adam Tooze[44] eloquently demonstrates in his great work on the Nazi economy and rise of militarism, the subject of defence is inextricably linked to the subjects of commercial capacity, commitment, and competence. From the perspectives of today, one might wish to offer themes of agility and adaptability into that mix. Second, as the retired defence minister and Chairman of the House of Commons Defence Select Committee, Lord Arbuthnot, puts it:[45]

> UK defence in the 21st Century is meaningless without industrial know-how and engagement. Without this, a commander is like a chef trying to cook dinner with no ingredients. He might have the ability to create a masterpiece but we can never know.

Notes

1 Interviewed in November 2014 by the authors.
2 Michael Clarke, 'Conclusion' in Adrian L Johnson (ed.), *Wars in Peace: British Military Operations since 1991* (London: RUSI, 2014).
3 For example, see Karl West and Oliver Shah, 'Hammond Prepares to Privatise MoD', *The Sunday Times*, 15 January 2012; Carola Hoyos, 'MoD Shifts Stance on Suppliers', *The Financial Times*, 1 February 2012.
4 See *DESider Magazine*, Issue 42, November 2011.
5 Author interview with a senior industrialist on 3 February 2012.
6 The author was briefed by a senior MoD official on 13 January 2012 that options for the reform of DE&S revolved around choices between some form of Government Owned, Contractor Operated (GOCO) organisational solution, the establishment of a trading fund, perhaps in partnership with a private sector company, or the creation of a Non-Departmental Public Body. For a number of officials and commentators, the debate should be concerned with not just the organisational solution proposed by any DE&S reform, but also its future spending powers and ability to raise capital.
7 See Ministry of Defence, *National Security Through Technology: Technology, Equipment and Support for UK Defence and Security*, CM8278 (London: The Stationery Office, February 2012). In the foreword, Peter Luff MP, Minister for Defence Equipment, Support and Technology reports that the government spends about £18 billion on defence from within the industrial base, most of this administered by DE&S.
8 *The Concise Oxford Dictionary*, Eighth Addition (Oxford: Clarenden Press, 1991).
9 See Dennis Lock (ed.), *The Gower Handbook of Management*, 3rd edition (Aldershot: Gower, 1992).
10 See Richard Lamming, *Beyond Partnership* (London: Prentice Hall, 1993).
11 See Mari Sako, *Prices, Quality and Trust: Inter-firm Relations in Britain and Japan* (Cambridge: CUP, 1992).
12 Author interview with MoD official, 13 January 2012.
13 Author interviews with former MoD officials and senior industrialists, December 2011.
14 See Paul Du Gay, *In Praise of Bureaucracy* (London: Sage, 2000).
15 See Chris Lonsdale, 'Post-contractual lock-in and the UK Private Finance Initiative' in *Public Administration* 83(1), 2005 pp. 67–88.
16 See John Louth, *A Low Dishonest Decade: Smart Acquisition and Defence Procurement into the New Millennium* (Cardiff, UWIC, 2010).
17 See Jean Froud, 'The Private Finance Initiative – Risk, Uncertainty and the State', in *Accounting, Organizations and Society* 28(6) (2003) pp. 567–89; A Coulson, 'Value for Money' in *Public Administration* 86(2), 2008, pp. 438–98.
18 See Lamming (1993), op. cit.
19 See K Kernaghan, 'Partnership and Public Administration: Conceptual and Practical Considerations' in *Canadian Public Administration* 36(1), 1993 pp. 60–5.
20 Author interview with a former MoD official, December 2011.
21 See Ministry of Defence, *The Strategic Defence Review*, CM3999 (London: The Stationery Office, 1998).
22 Sourced from Ministry of Defence, *The Smart Acquisition Handbook* (Edition 4, 2002).
23 Sourced from Ministry of Defence, *The Smart Acquisition Handbook* (Edition, 5, 2004).
24 See the LTPA website, run by QinetiQ, at: www.ltpa.co.uk.index.asp.
25 The LTPA capabilities and range of services are explained at: www.ltpa.co.uk/capabilities/index.asp.
26 Author interview with a former MoD official, December 2011. The design objective of the agreement is also repeated on the LTPA website at: www.ltpa.co.uk/about/ltpa_explained.asp.

27 Author interviews with former MoD officials and senior industrialists, December 2011.
28 For a broad and accessible overview of strategic management principles see, for example, Barry Witcher and Vinh Sum Chau, *Strategic Management, Principles and Practice* (Andover: Cengage, 2010).
29 See Ministry of Defence, *Defence Industrial Strategy*, CM6697 (London: The Stationery Office, 2005).
30 See Steve Wadey, 'Progressing Team Complex Weapons' in *RUSI Defence Systems*, June 2010.
31 Discussed in a background briefing to RUSI staff members on 4 November 2011.
32 A senior military official from MoD informed the author in September 2010 that the CBRN Key Strategic Partnership programme was even at that time being perceived as a test-case for broader DE&S reform.
33 The author was, for a short time, involved in the development of the assessment phase programme at its inception.
34 Joint CBRN KSP Team Communication: Issue No 1, November 2010.
35 A senior project manager from DE&S presented at the Industry Day and reported that 'partnering' and 'partnership' were synonymous concepts for the CBRN KSP team members.
36 An extensive and detailed Concept of Analysis was agreed within MoD during November 2009.
37 The author talked with both MoD and industry members of the Joint Delivery Team in December 2011 and January 2012.
38 See John Louth and Peter Quentin, 'Making the Whole Force Concept a Reality,' *RUSI Briefing Paper*, November 2014.
39 Ministry of Defence, 'Operations in Iraq – Lessons for the Future' (London, 2003), p. 6.
40 Andrew Higginson, 'Contractor Support to Operations (CSO) – Proactive or Reactive Support?', *RUSI Defence Systems* (October 2010), p. 16. This argument was also made in Henrik Heidenkamp, John Louth and Trevor Taylor, *The Defence Industrial Ecosystem: Delivering Security in an Uncertain World* (London: Royal United Services Institute, 2011).
41 HM Treasury, *Budget 2010*, HC61 (London: June 2010).
42 Ministry of Defence, Air Commodore Dan Hill, 'What is the "Whole Force" Concept?', 21 March 2011.
43 In a private interview with the author in September 2016, a senior UK industrialist from a defence prime contractor stated that the company had planned for 1,000 sponsored reserves under an enabling contract with the MoD. To date, only 49 such posts had been reclassified as sponsored reserves.
44 Adam Tooze, *The Wages of Destruction: The Making and Breaking of the Nazi Economy* (London: Penguin Books, 2006).
45 Lord Arbuthnot speaking with the authors in June 2016.

Bibliography

Clarke, M (2014), 'Conclusion' in Adrian L Johnson (ed.), *Wars in Peace: British Military Operations since 1991* (London: RUSI).
Coulson, A (2008), 'Value for Money' in *Public Administration* 86(2), pp. 438–98.
DESider Magazine, Issue 42, November 2011.
Froud, J (2003), 'The Private Finance Initiative – Risk, Uncertainty and the State', in *Accounting, Organizations and Society* 28(6), pp. 567–89.
Du Gay, P (2000), *In Praise of Bureaucracy* (London: Sage).

Heidenkamp, H, Louth, J and Taylor, T (2011), *The Defence Industrial Ecosystem: Delivering Security in an Uncertain World* (London: Royal United Services Institute).

Higginson, A (2010), 'Contractor Support to Operations (CSO) – Proactive or Reactive Support?', *RUSI Defence Systems* (October 2010).

HM Treasury (2010), *Budget 2010*, HC61 (London: June 2010).

Hoyos, C (2012), 'MoD Shifts Stance on Suppliers', *The Financial Times*, 1 February 2012.

Joint CBRN KSP Team Communication: Issue No 1, November 2010.

Kernaghan, K (1993), 'Partnership and Public Administration: Conceptual and Practical Considerations' in *Canadian Public Administration* 36(1), pp. 60–5.

Lamming, R (1993), *Beyond Partnership* (London: Prentice Hall).

Lock, D (ed, 1992), *The Gower Handbook of Management*, 3rd Edition (Aldershot: Gower).

Lonsdale, C (2005), 'Post-contractual lock-in and the UK Private Finance Initiative' in *Public Administration* 83(1), pp. 67–88.

Louth, J (2010), *A Low Dishonest Decade: Smart Acquisition and Defence Procurement into the New Millennium* (Cardiff, UWIC).

Louth, J and Quentin, P (2014), 'Making the Whole Force Concept a Reality,' *RUSI Briefing Paper*, November 2014.

LTPA website, run by QinetiQ, at: www.ltpa.co.uk.

Ministry of Defence (1998), *The Strategic Defence Review*, CM3999 (London: The Stationery Office).

Ministry of Defence (2002), *The Smart Acquisition Handbook* (Edition 4).

Ministry of Defence (2003), 'Operations in Iraq – Lessons for the Future' (London).

Ministry of Defence (2004), *The Smart Acquisition Handbook* (Edition 5).

Ministry of Defence (2005), *Defence Industrial Strategy*, CM6697 (London: The Stationery Office).

Ministry of Defence (2011), Air Commodore Dan Hill, 'What is the "Whole Force" Concept?', presentation given at RUSI, 21 March 2011.

Ministry of Defence (2012), *National Security Through Technology: Technology, Equipment and Support for UK Defence and Security*, CM8278 (London: The Stationery Office, February 2012).

Sako, M (1992), *Prices, Quality and Trust: Inter-firm Relations in Britain and Japan* (Cambridge: CUP).

The Concise Oxford Dictionary, Eighth Addition (Oxford: Clarenden Press, 1991).

Tooze, A (2006), *The Wages of Destruction: The Making and Breaking of the Nazi Economy* (London: Penguin Books).

Wadey, S (2010), 'Progressing Team Complex Weapons' in *RUSI Defence Systems*, June 2010.

West, K and Shah, O (2012), 'Hammond Prepares to Privatise MoD', *The Sunday Times*, 15 January 2012.

Witcher, B and Chau, VS (2010) *Strategic Management, Principles and Practice* (Andover: Cengage).

10   Defence as the military

Introduction

Within this chapter we turn to the more conventional idea of viewing defence as being principally the property and concern of the military. We discuss the size and shape of the UK armed forces, from the baseline of the country's military condition in 2010 – the year the Coalition government, between the Conservative Party and Liberal Democrats, was formed. From there we describe the defence cuts made to the British military during that parliament, between 2010 and 2015, and discuss the supposed recapitalisation of the military since then under the banner of Joint Force 2025.

We will see, perhaps, that the modern British military is an amalgam of its history, the ambitions placed on it by its own senior leaders and others, the range of threats it faces, and, given the chapters that precede this, the complexities of integration across multiple private and public-sector partners, and understandings, that comprise modern UK defence. The service woman or man is revealed as not just a fighter, but a manager, a technologist, an implementer of social and industrial policy, an exporter, partner, and someone well-versed in the skills of the storyteller. The modern military professional is, we suggest, a different person from that which the casual reader might have expected to encounter. How can it be otherwise given the previous pages?

Chapter objectives

By the end of this chapter the reader will understand:

1   The role of the military within the UK defence system;
2   The organisation of the military into the three armed services and the role of Joint Forces Command;
3   The transformation of the military as it shapes for a posture labelled as Joint Force 2025;
4   The challenges faced by the military.

Chapter structure

This chapter starts with a discussion of the structure and disposition of UK military forces in 2010, by way of a baseline. It discusses the roles assigned to these forces by the UK government in support of our defence posture and treaty obligations. Thereafter, the chapter considers the reform of the military during the Coalition government of 2010–2015, its reduction in size, and the intended recapitalisation to form Joint Force 2025. The chapter concludes with a discussion of the problems facing the British military.

Sir Julian Brazier, a former defence minister during the Coalition government of 2010–2015, has described a tension at the heart of the military condition:

> There is a paradox in the relationship between the military and our society. People feel less secure yet, despite this, less engaged with their military than perhaps ever before.[1]

As we have discussed in previous chapters, the risks of peer-on-peer conflicts and the potential shocks that could be caused by terrorist outrages within our borders do appear substantial to the citizen, even overwhelming. By way of a highly subjective experiment, at a conference on future trends in Oxford in July 2017,[2] the authors asked for a show of hands of an audience of 200 delegates whether people felt more or less secure, in a defence sense, in 2017, than they felt in the mid-1980s. It was virtually unanimous that people felt less secure today than in the years before, despite the twentieth century being the period of total war, cold war, and the development of doctrines such as nuclear mutually assured destruction.[3]

The data points to a different truth. In antiquity, human violence generally caused 15 per cent of violent deaths. During the twentieth century it averages just 5 per cent of all deaths whilst, in the twenty-first century, so far at least, human violence accounts for just 1 per cent of deaths. In 2012, 56 million people died. Of this number, 620,000 human beings died violently, with 500,000 deaths associated with crime (a high proportion being women killed at the hands of a male partner, relative, or associate) and only 120,000 deaths associated with conventional notions of war or terrorism. This is a tiny number, in contrast significantly to suicide which claimed 800,000 lives in the same year.[4] Yet, despite the data, people *feel* insecure and yearn to be protected, including presumably by their military. The military instrument, therefore, has to be developed and matured to address feelings as well as facts and to be prepared for risks, uncertainty, and the contingent event rather than the certainty of a known state enemy with clear intent. This, of course, has to be done without adding to international tension or causing a political or military miscalculation. What Winston Churchill wrote in 1948 remains an ominous warning almost three-quarters of a century later:

> [Mankind] has got into his hands for the first time the tools by which it can unfailingly accomplish its own extermination ... Death stands at attention,

obedient, expectant, ready if called on to pulverize, without hope of repair, what is left of civilization. He awaits only the word of command. He awaits it from a frail, bewildered being, long his victim, now – for one occasion only – his Master.[5]

Military leaders, planners, and their soldiers, airmen and women and sailors, and the politicians that commit them, have a range of very difficult calculations to make.

The role of the military

It would be more accurate to describe the UK military as possessing a variety of roles. Its principal objective, though, is to keep the British homeland safe and secure, historically from other rival state actors. So, a military has to deter, coerce a rival power to its national will, or defeat that rival power on the battlefield. In addition, states now ask their militaries to undertake aid to civil authorities following a natural disaster or during a national economic dispute. Our society also wants the military to help the government and commercial partners sell arms and defence systems to friendly states, even if these are not democracies. These are very different purposes, of course, and require quite different characteristics and skill-sets.

Deterrence

In the literature relating to international relations, deterrence equals diplomacy and is not a natural military concept.[6] The argument is that the military professional's primary test is typically undertaken on the battlefield through the controlled use of violence utilising weapons, technologies, and proven techniques. Deterrence, in contrast, is based on weapons and technologies, but the practice of deterrence is the threat of force rather than its actual use. This manipulation of a potential enemy through the threat of force is one aspect of diplomacy and the practice of bargaining for peaceful coexistence between powerful states.

Deterrence is exercised both though conventional military forces and, in the case of the UK, through its nuclear capabilities – the continuous, at sea deterrence offered by the four trident strategic submarines and its successor programme. The enduring principle is that an adversary will suffer significant pain and disruption – not to say total destruction – if it attacks the homeland of the UK, its forces, or the territories and assets of its allies.

Coercion

The idea of coercion is tied to the concept of deterrence. If a state or non-state actor is aware that the UK has the capability and the will to substantially hurt it, through the use of conventional or (exceptionally) nuclear forces, the argument seems to be that this adversary will bend towards the will of the UK. Operations in Afghanistan and Iraq seem to contradict this classic theory of international politics but, no doubt, the ability and will to inflict multiple casualties and massive destruction can be exploited diplomatically *before* war breaks out. Modern technologies, destructive power, and reach, have all enhanced the significance of the UK's military potentially being used as a tool of influence and coercion rather than conquest or annihilation.[7]

Victory

If, though, the UK's armed forces have to be committed on operations or in a war of national survival, the imperative is for those forces to be victorious.[8] To do so, an enemy's political, societal, economic, and operational will, and ability to fight has to be reduced or destroyed. The recognised levels of warfare from which planning, command, and operations are derived are the *Grand Strategic*, the *Military Strategic*, the *Operational*, and the *Tactical*. 'The essence of planning at each level is to balance the desired end, the way in which it is to be achieved, and the adequate means of achieving it.'[9] Grand strategy is the application of national resources to achieve national policy objectives and those of allies or coalitions. Military strategy, not surprisingly, is the use of military resources to help achieve stated grand strategic aims or effects. The operational level is concerned with the direction of those military resources to meet strategic military objectives, whilst the tactical level is concerned with activities undertaken by the military to achieve operational objectives.

Thus, by way of example, in the First Gulf War in 1991, the grand strategic level was set by the prime minister to join the coalition to liberate Kuwait. This decision was taken in consultation with the United Nations, which saw the passing of United Nations' Security Council Resolution 678 mandating the use of military force if the Iraqi military did not withdraw from Kuwait. The military strategic level saw the US-led senior coalition leadership set grand strategic targets and objectives within the military sphere of operations. The Joint Coalition Headquarters, under the command of US General Norman Schwarzkopf, exercised operational command and control, whilst the tactical level of the plan was exercised by various component commanders, such as the commanders of Royal Naval Task Force and the 1st (United Kingdom) Armoured Division.

Military aid to the civil authorities

Military aid to civil ministries is typically the use of military personnel for non-military governmental tasks such as fire-fighting, logistics support to the farming sector, or the provision of ambulance services when these civil functions are threatened by natural disaster or industrial disputes. In contrast, military aid to the civil power is traditionally used to return or maintain law and order beyond the capabilities of the civil authority at that moment. This is undertaken at the request of the civil power and the military usually remain under the operational control of that civil power. If, exceptionally, control of an incident is passed from the civil power to the military, it is returned to the civil power at the earliest opportunity. An example of this role could be counter-insurgency, counter terrorism, or explosive ordnance disposal.

The UK military of 2010

With this as background, the military has tended to classify the types of operations it could be asked to undertake under the following headings:

- Combat. This can be subdivided, in turn, into high intensity warfare with a clearly defined state enemy; armed intervention involving the entry of combat forces into the territory or area of jurisdiction of another state; counter-insurgency operations against an armed rising or terrorist affiliates; counter terrorist operations within the UK under the direction of the civil authority and subject to domestic law.
- Deterrence. Conventional forces contribute by demonstrating effectiveness on operations and a level of readiness for a wide range of operations. Nuclear forces are perceived as the optimum or strategic level of deterrence.
- Support to diplomacy. The use of military personnel to assist diplomatic efforts internationally either in terms of crisis support or routine support. An example of the former could be the evacuation of personnel following a natural disaster, or the provision of humanitarian aid. The most prominent example of routine support is the use of Defence Attaches – military officers seconded to the Foreign and Commonwealth Office – within UK embassies and missions overseas.
- Peacekeeping. These operations are authorised by the United Nations, or an equivalent regional body, and carried out with the general consent of the belligerent forces and civil populations.
- Peace-making. Enforcement operations can be carried out under the authority of the United Nations' Security Council to force peace through military intervention.
- Military Home defence. This comprises the activities necessary to protect and preserve the functions of state and critical national infrastructure in times of crisis or war.
- Military aid to the civil authorities. As discussed, this is aid to the civil communities, ministries, or civil power.
- Monitoring compliance with Arms Control Treaties. Treaties such as the limit on Conventional Forces in Europe (CFE) can be monitored through the use of military compliance officers.
- Public and ceremonial duties. The UK military plays a role in national life through ceremony and public events.[10]

Form and structure

To undertake these multiple roles – characterised by us as principally concerned with deterrence, coercion, and military intervention for victory, the UK military in 2010 was organised in the following manner.

The regular armed forces of the UK comprised 187,210 personnel, not including 3,600 Gurkhas and 2,040 Full-time Reserve Service personnel.[11] This was made-up of:

- Royal Navy: 38,160 including 7,500 Royal Marines;
- British Army: 105,750 soldiers;
- Royal Air Force: 43,290.

In addition, at this time there were 40,780 Volunteer Reserves and 80,970 Ministry of Defence civil servants. These women and men populated their respective services to generate the following force elements:

Royal Navy:

- Four nuclear powered ballistic missile submarines – forming the UK strategic nuclear deterrent;
- Nine nuclear powered attack submarines;
- Three aircraft carriers, fielding a mix of fixed wing harrier aircraft and helicopters;
- One helicopter carrier;
- Two amphibious assault ships;
- Eight destroyers;
- Eighteen frigates;
- Sixteen minehunters and minesweepers;
- Five ocean survey vessels;
- One Antarctic patrol ship;
- Four patrol vessels;
- Sixteen patrol craft and fishery protection craft;
- Fleet Support Ships manned by the Royal Fleet Auxiliary personnel:
  - 2 × fast fleet tankers;
  - 2 × small fleet tankers;
  - 4 × support tankers;
  - 4 × replenishment ships;
  - 1 × aviation training ship;
  - 1 × forward repair ship;
  - 4 × landing ships.
- Royal Marines:
  - 1 × Commando Brigade Headquarters;
  - 3 × Royal Marine Commando battalions;

- 3 × Commando Assault helicopter squadrons;
- 1 × Commando Light Helicopter squadron;
- 1 × Commando regiment Royal Artillery;
- 1 × Commando squadron Royal Engineers;
- 1 × Commando Logistics regiment;
- 1 × Commando Assault Group (Landing Craft);
- 1 × Fleet Protection Group.

British Army:

- One Corps Headquarters (Allied Rapid Reaction Corps);
- Two Divisional Headquarters, one in Germany and one in the UK;
- Five non-deployable Headquarters in the UK;
- Eight deployable Brigade Headquarters;
- Ten Regional Brigade Headquarters;
- Ten armoured regiments;
- Thirty-six infantry battalions;
- Fifteen artillery regiments;
- Eleven engineering regiments;
- Twelve signal regiments;
- Seven equipment support battalions;
- Seventeen logistics regiments;
- Eight medical regiments.

Royal Air Force:

- Five strike/attack squadrons;
- Two offensive support squadrons;
- Five reconnaissance squadrons;
- Four defence interceptor squadrons;
- Two maritime patrol squadrons;
- Two airborne early warning squadrons;
- One intelligence, surveillance, target acquisition, and reconnaissance (ISTAR) squadron;
- Eight transport/tanker squadrons;
- Two helicopter search and rescue squadrons;
- Six ground defence squadrons.

In addition, the Permanent Joint Headquarters (PJHQ) at Northwood plays a key role within UK defence and possessed in 2010 a number of joint assets, as described below. When directed by the Chief of the Defence Staff, PJHQ is responsible for planning and executing joint and multinational operations led by the UK. Importantly, it is also responsible for exercising operational command over UK forces assigned to multinational operations led by others. It is commanded by the Chief of Joint Operations and in 2010 could field the following forces in addition to forces generated from the single service commanders:

- Joint Force Harrier, comprising two squadrons from the Royal Navy and two from the Royal Air Force.
- Joint Helicopter Command: This possessed four Royal Navy helicopter squadrons, six Army aviation regiments, and seven RAF helicopter squadrons.
- Joint Nuclear, Biological, and Chemical regiment.
- Joint Special Forces Group: one Special Air Service (SAS) regiment, two Volunteer Reserve SAS regiments, four Royal Marines Special Boat Service (SBS) squadrons, one special reconnaissance regiment, and one Special Forces Support Group.

These combined forces at our benchmark year in 2010 are substantial and impressive. The age of austerity, the perceived need for public spending savings, and the sense that military force could not prevail in places like Afghanistan and Iraq would collude to dent this UK military mass. Welcome to the 2010 Strategic Defence and Security Review of 2010.

The Strategic Defence and Security Review 2010

The UK's publication of its National Security Strategy and the Strategic Defence and Security Review (SDSR)[12] in 2010 confirmed that the UK budget for its national defence had to contract by about 7.5 per cent to 2015, and that military capabilities would need to be reduced and rationalised. This led to platforms such as the Harrier aircraft and the Nimrod MRA4 anti-submarine warfare aircraft being removed immediately and completely from the UK battle plan. The UK Parliament Defence Select Committee in 2011 heavily criticised UK government policy for defence, claiming that UK forces would in future struggle to meet its commitments. Moreover, the Committee asserted that capability gaps would emerge in the short- to mid-term, such as Carrier-Strike capabilities and maritime patrol, and that in the mid- to long-term the UK would no longer be a full-spectrum defence capable nation.[13] The chairman of the Defence Select Committee, James Arbuthnot MP, went so far as to say that SDSR:[14]

> ... is a clear example of the need for savings overriding the strategic security of the UK and the capability needs of the Armed Forces. The Government needs to outline its plans to manage the gap left by the loss of these capabilities.

Moreover, a number of National Audit Office reports[15] argued that substantial systemic and behavioural issues around the management of UK defence equipment, relationships, and personnel had bequeathed a cycle of unrealistic requirements setting, planning, competition policies, budgeting, and contractual practices. This, in turn, led to substantial cost overruns, delays to equipment in-service dates, and the necessity for the MoD to find significant short-term budgetary savings. Moreover, according to the MoD's senior political leadership, at the beginning of the Coalition government's term of office the UK had a hole in its defence budget to the tune of £51 billion. This was comprised of an 'unfunded liability' of £38 billion recorded in the SDSR, a further £8 billion to potentially renew the nuclear deterrent in some form, and a £5.5 billion revaluation of the core equipment programme ordered by Bernard Gray, the Chief of Defence Materiel: a ten-year funding gap of £51 billion from 2010/2011 to 2020/2021.[16] This funding gap, as well as the uncertainties associated with future budgetary constraints, the need for economies, changing requirements, an operational drawdown from Afghanistan by 2014, and the ongoing organisational restructuring of the military itself,[17] dominated the UK defence and national security environment.

In addition, despite a history of indigenous defence industrial capabilities stretching back over epochs, there was a growing unease in relation to defence industrial policy under the UK Coalition government. As stated previously, it was UK government policy to have a complete range of defence capabilities to meet seven key strategic tasks[18] ranging from the defence of the UK and overseas sovereign territories, through an ability to project power via expeditionary interventions, to the support to civil authority in response to natural, and other, emergencies. Likewise, it was government's responsibility to align resources to required capabilities in order to meet these goals. If resources are not there, or savings are required in pursuit of other considered government policies that have a higher priority than defence, then the defence capability demanded has to be reduced. It seemed a legitimate concern at the time that the UK government, like those before it, had failed to understand, or chose to ignore, the causal relationship between capability demanded and resources needed.

The White Paper on Technology, Equipment and Support[19] hinted at an ambition for much of the UK's military hardware needs to be procured from 'off the shelf' (OTS). For advocates, this policy was perceived as a cheaper and less-risky option than researching and developing new defence equipment within the UK. However, most OTS packages would inevitably mean 'off-shore' purchases leading to questions around intellectual property (IP) transfers. Without fully owning or even understanding the IP, British military practitioners would not be able to completely exploit the equipment in question; a most unsatisfactory scenario. Moreover, the costs of through-life upgrades of military OTS equipment can be prohibitive, undermining any initial value-for-money assessments. Also, in times of national crisis, it is very difficult to possess a surge-capability within a national industrial base if that national industry base no longer exists. This represents, of course, a probable consequence of reliance on OTS procurement.

This sense of the early stages of reformation of government defence industrial policy was highly significant and remains so. In the UK, nine companies were paid more than £500 million by the MoD at the time the Coalition government came to power. For the same year, 2010, seven companies were paid between

£250 million and £500 million. Thirteen companies were paid between £100 million and £250 million, whilst 22 companies were paid £50 million to £100 million. Eighty per cent of these companies were British listed or private businesses.[20] These are sizeable sums that many of the companies in question rely upon in order for them to remain viable commercial entities. The twin challenges of budgetary constraints and OTS purchases may well drive some companies from the market in the mid-long term. Indeed, one senior industrialist commented privately that his company could substantially reduce operations in the UK, perhaps even relocating and listing elsewhere, if a more involved and proactive UK defence industrial policy failed to emerge in the months and years ahead.[21]

Joint Force 2025 – a recapitalisation of the UK military

Strategic Defence and Security Review 2015

The UK Strategic Defence and Security Review in November 2015 was a mix of a National Security Strategy and a conventional Defence and Security Review.[22] The national plan identified the government's intention to meet the security challenges identified across government, setting the mandate for the UK Armed Forces over the coming ten years in terms of planning, capability requirements, equipment provision, priorities, likely missions, and manpower ambitions and limitations. One of the important outcomes of the new policy was an articulation of what the government would aim to achieve in terms of the shape of Britain's armed intervention forces in ten years' time. This was to sit alongside core activities of the British military, including the provision of the nuclear deterrent, air defence of the UK mainland, defence of overseas territories, international defence obligations, and support to civil authorities in case of a crisis at home.

The intervention force, named Joint Force 2025, was the outcome of a complex process informed by an amalgam of risk and threat analyses, inherited commitments and obligations, manifesto commitments, available and projected resources, assumptions on cost trends, and the potential for savings. It was a reflection of the ambition of the British government to craft an understanding of the threats to defence and security of the UK and allies within a rigid financial budget.

The key components of SDSR 2015, to yield Joint Force 2025, in terms of force elements and equipment packages, were to be as follows:[23]

Royal Navy:

- Four nuclear powered ballistic missile submarines – forming the UK strategic nuclear deterrent;
- Seven nuclear powered attack submarines;
- Two aircraft carriers, fielding the Lightning II, Joint Strike Fighter F35b and, possibly, a mix of helicopters;
- 3 Commando Brigade, with two landing platform docks and three landing ships;

- Nineteen frigates and destroyers;
- Twelve minehunters;
- Three ocean survey vessels;
- One Antarctic patrol ship;
- Up to six patrol vessels;
- Four Merlin Mk 2 helicopter squadrons and two Wildcat squadrons;
- Fleet Support Ships manned by the Royal Fleet Auxiliary personnel:

  - 6 × Fleet tankers;
  - 3 × Fleet Solid Support Ships.

British Army:

- Two Armoured Infantry Brigades;
- Six Infantry Brigades or eighteen infantry regiments;
- Two Strike Brigades;
- Sixteen Air Assault Brigade;
- Four Apache squadrons, four Wildcat squadrons and three Watchkeeper batteries;
- 77 Brigade (Information Warfare);
- One Intelligence, Surveillance, and Reconnaissance Brigade.

Royal Air Force:

- Seven Typhoon squadrons;
- Over 20 Remotely Piloted Protector aircraft;
- Two Lightning II, F35 squadrons;
- Nine P8 Maritime patrol aircraft;
- Up to 27 early warning and reconnaissance aircraft, with a mix of Rivet Joint, Shadow, and E-3D entry aircraft;
- Fourteen Voyager aircraft;
- Eight C17 aircraft;
- Twenty-two A400M Atlas and 14 C130J Hercules aircraft;
- Six Force Protection Wings;
- Two Puma squadrons, three Chinook squadrons, and two Merlin squadrons.

In terms of headcount, the military full-time trained strength targets for 2010 were:

- Royal Navy and Royal Marines: 30,450 personnel;
- British Army: 82,000 personnel;
- Royal Air Force: 31,750 personnel;
- These were to be supplemented by a volunteer reserve trained strength in 2020 of 35,060 Reservists, made up of 3,100 part-time sailors, 30,100 soldiers, and 1,860 airmen and women.

So, between 2010 and the aspirations around Joint Force 2025, the Royal Navy and Royal Marines full-time strength has shrunk from 38,160 to 30,450, or by 20 per cent. The British Army has shrunk from 105,750 to 82,000 – by 23,750 soldiers or 22.5 per cent. And the Royal Air Force has lost 11,540 personnel (from 43,290 to 31,750) or 27 per cent of its force. In more than one way, Sir Michael Fallon, Secretary of State for Defence, is entitled to state:

> SDSR 2015 opens a new chapter for defence. It shows how we will protect our people and interests in the coming years. In particular, it sets out plans to tackle the threat of extremism and state aggression, to strengthen the rules-based international system and to increase our prosperity.[24]

All with a smaller military, of course. For example, between our base year of 2010 and Joint Force 2025, the infantry regiments of the British Army have shrunk in number from 36 to 18 – a 50 per cent reduction. What this represents – in terms of ambition and realism – we now come on to discuss.

Joint Force 2025 – ambition versus realism

It is observable that, for many years now, successive British governments have announced their plans for securing the defence of the UK and its peoples at the start of their terms of office, usually - but not always - founded on their perceptions of the strategic security environment and the threats faced by the nation. This seems logical and sensible. For example, the 2010 SDSR outlined new defence planning assumptions, articulating what the government was planning for in terms of scale, endurance, how these operations were to be undertaken, and how far away they might be from our shores.[25]

In 2015, this new policy broke with the previous model of delineating deployment models (between standing commitments, stabilisation, and intervention operations). Instead a large, integrated, intervention force (JF2025) was to be generated. If it was not required, from this standing posture the force could be used for multiple small-scale activities. The weight of effort seemed to have shifted from a preparation for small-scale, short-term 'complex' interventions and towards a focus on the best efforts of the UK armed forces in terms of capability, commitment, and scale.

The then prime minister, David Cameron, stated that:

> We will be able to deploy a larger force more quickly. By 2025, this highly capable expeditionary force of around 50,000 (compared with the planning assumption of around 30,000 Future Force 2020, its predecessor) would include:
>
> * A maritime task group centred on a Queen Elizabeth Class aircraft carrier with F35 Lightning combat aircraft.
> * A land division with three brigades including a new Strike Force.

- An air group of combat, transport and surveillance aircraft.
- A Special Forces task group.[26]

The Secretary of State for Defence added little to this announcement about JF2025 in the subsequent period. His contribution at the time was limited to additional JF2025 fact sheets accompanying the publication of SDSR 15. Rather, he restated:

> We're establishing a new Joint Force 2025 with a raft of cutting-edge capability. This includes new carriers, hunter killer subs, frigates and fighter jets as well as multi-mission aircraft capable of maritime patrol, strike brigades and armoured vehicles, double the number of Unmanned Aerial Vehicles and more Special Forces capabilities. We're doing this by working more closely with our allies, harnessing the power of innovation, and continually improving our productivity.[27]

There was little accompanying explanation of what the force was designed to do in conceptual terms, or how it would be employed, except that it was a balanced force capable of meeting challenges from terrorism to state actors. By comparison, previous reviews had outlined the concept underpinning the force posture.

SDSR 15 outlined an aspiration for greater cross-government participation in defence and security capabilities and missions, specifically in the area of 'integrated' command and control.[28] The policy stated that integrating the intelligence agencies, security services, and police forces more effectively was the way in to deliver more efficient and time-sensitive decision-making that could be made in order to exploit capabilities better, and use national power more coherently. Subsequently, the government established a Full Spectrum unit/organisation in the Cabinet Office, and another in the FCO with responsibility for cohering HMG approaches. Yet the very fact that there are two organisations and not simply one highlights a lack of coherence that the 2015 SDSR sought to overcome.

Aside from the establishment of these organisations, it appears that little has been done internally within government to develop thinking in these areas or to ascertain whether the planned changes to command and control have brought about the efficiencies imagined and the effectiveness demanded. Changes to the national security command structure have not been evident, nor has a different version of political-military command and control emerged.

Joint Force 2025

The force outlined in the SDSR 15 was an articulation of the integrated, modern British warfare tool that could have been capable of being 'decisive' on a modern battlefield against a peer adversary as part of an alliance operation, but also capable of contributing to operations against a range of other actors. It was seen as being able to provide capabilities for standing commitments, and of integrating all the levers of British power, cognisant of all the lessons of warfare and acknowledging Britain's vital interests.

The Review aligned the UK approach with that of the United States in terms of a focus on innovation,[29] but was underpinned by an internal Straw-man paper titled 'Warfare in the Information Age' produced in December 2014 by General Sir Richard Barrons, then commander of Joint Forces Command (and subsequently reproduced in a developed form in the RUSI Journal).[30] Both the original paper, and subsequently the SDSR 15, argued that a rebalance into intelligence, surveillance, and reconnaissance was required across the armed forces, and was the central argument driving platform purchases announced within SDSR 15. These included P8 Poseidon Maritime Patrol Aircraft, AH64 Apache attack helicopter upgrades, new drones, greater focus and investment into airborne Intelligence and Surveillance platforms (Sentinel programme), new frigates, upgrades to the Challenger II Main Battle Tank, and a new fleet of fighting vehicles for the British Army named Ajax. JF2025 was purported to be a significant departure from the post-Cold War downsizing and salami slicing of capability that the British forces had been subject to, albeit one that was looking for a mission or role.

This posture was described by Michael Clarke as a Strategic Raiding force.[31] To the casual observer of military tactics, this meant a force capable of small, short, and fast engagements at distance, but not one capable of delivering battle-winning forces at scale or with durable fighting power – let alone against a near peer state adversary. In short, was JF2025 useful or just an odd mix of narratives, concepts, and capabilities?

The global context

Whilst the global security context has not fundamentally changed since the SDSR, there has been a noticeable evolution in trends and patterns. The threats from Russia, China, terrorism, and migration were a core part of the construction of the 2015 SDSR process. Each of these challenges has mutated in form since 2015 and will, no doubt, continue to do so. Terrorist attacks in continental Europe in November and December 2015 continued through 2016 and into 2017, but not in a way that seemed to shock other European capitals. Rather, it seems that these attacks, along with increased migration flows into mainland Europe, have increased calls for an increase in border controls within Europe and greater calls for military intervention to counter the impacts of instability in North Africa and Syria.

Chinese actions in occupying and expanding military capabilities in the China Seas became clearer during early 2016, threatening both free trade and, more broadly, international normative behaviours. Russian military and deniable activity in the Baltics, Balkans, Ukraine, Georgia, and the North Atlantic served to undermine both NATO cohesion and regional stability. The Russian actions in Syria in 2016 took an increasingly menacing tone, with their air and naval strikes in support of the Assad regime, and in increased basing of Russian military assets on the shores of the Mediterranean Sea.

In July 2016, Sir John Chilcot's Iraq Inquiry was published, making far-reaching conclusions about the failings of both the British political and military machines in the conduct of conflict. In June, the British referendum on EU membership delivered a clear decision to leave the European Union unpicking many of the supporting assumptions contained within SDSR 15 only six months previously. Added to this, in November 2016, the election of Donald Trump as President of the United States occurred, bringing with it potentially significant implications for Britain's defence posture and relationship with its key strategic ally. Within this period also, election results in Moldova and Bulgaria put pro-Russian governments into power.

Such events undermined the narrative arc that formed the foundations of the Joint Force 2025. Could the nature of traditional alliances be really relied upon in times of crisis? Would the UK-US 'Special Relationship' stand the new tests of two highly transactional leaders – as opposed to a pairing that, according to popular culture at least, shared common values? Would a dependence on burden-sharing between European partners work, given the apparent divergence of security priorities between states? We don't know, but the condition of UK defence and our security within the word is, in part at least, dependent upon these answers.

Joint Force 2025 – deterrence, coercion, victory?

The strategic raiding force of JF2025, as articulated in 2015 by Cameron, is certainly an upgrade on the messages of SDSR 2010, which foretold a period of strategic decline with substantial cuts being made to the British military. Indeed, the case could be made that, with more spending on people and equipment, and with a new defence innovation fund, the British armed forces were being recapitalised and that JF2025 represented a new dawn for the military. This would be precipitous in our judgement. Rather, the complexities of the world and the rather rigid platforms upon which JF2025 sits, suggest that a more coherent case for JF2025 needs to be made in the context of a force configured for deterrence, coercion, or military victory.

Moreover, a sense of the new security environment has become clearer, in the eyes of decision-makers and military planners, since SDSR 2015 was announced. The emergence of an understanding of how Russian and Chinese forces, for example, are constructed to conduct regional wars based on spheres of influence as opposed to the West's own development of interventionist, at distance warfare, provide clearer indications that the conduct of war has changed significantly since JF2025 was announced. A strategic raiding force would seem to have less utility against Russian or Chinese ambitions than one designed for a short-sharp intervention, leaving the broader international community to focus on stabilisation and recovery. This does not marry well with the aspirations outlined in the 2015 SDSR either, which focused on an ability to intervene, deter aggressors, and stabilise with equal priority.

It looks increasingly likely that JF2025 was not designed as a strategic raiding force but is being marketed as more than the sum of its parts. The UK military as a critical component of UK defence seems less sure of its role in the future order of battle than could be thought desirable.

Conclusion

This chapter has considered the subject of UK defence from the perspective of the military – that is the armed forces of the Royal Navy, British Army, and Royal Air Force. We discussed the structure and composition of the military, as it was comprised in 2010, and demonstrated the remarkable drawdown in capability between 2010 and 2015, as the Coalition government pruned the defence budget by 7.5 per cent. Despite the supposed recapitalisation of the military through SDSR 2015 and the refreshed defence posture associated with JF2025, the British military is significantly smaller in terms of manpower and force elements in 2020 than it appeared in 2010.

The chapter also considered the multiple roles performed by the military, from deterrence and coercion of our enemies through to the generation of victory on the battlefield. Coupled to these core roles is the requirement, on request, to provide assistance and aid to the civil authorities and civil society.

The military is at the heart of the UK's defence effort, as these pages have shown. But that defence 'practice' is a complicated and multi-faceted endeavour involving more than just the defence of borders and peoples. In the context of the book as a whole, the forces of the Royal Navy, British Army, and Royal Air Force are but one large component within a complicated and amorphous public–private partnership that generates the Defence Extended Enterprise. It is to this subject that we now turn in the following, concluding chapter.

Notes

1 Sir Julian Brazier, speaking to the authors on 26 July 2017.
2 Ordnance Survey Conference – the Cambridge Futures' Conference – held at Keble College, Oxford on 3 July 2017.
3 During the height of the Cold War between the NATO alliance and Warsaw Pact countries, the deterrence associated with a strategic nuclear exchange was characterised as mutually assured destruction, thereby asserting that no state or bloc could emerge victorious from such an exchange. See John Spanier, *Games Nations Play* (New York: Holt, Rinehart and Winston, 1984) pp. 159–60.
4 See Stephen Pinker, *The Better Angels of our Nature: Why Violence has Declined* (New York: Viking, 2011) and the World Health Organisation, 'Global Health Observatory Data Repository', 2012. Both data sources quoted in Yuval Noah Harari, *Homo Deus: A Brief History of Tomorrow* (London: Harvill Secker, 2015), pp. 14–15.
5 Winston S Churchill, *The Second World War, The Gathering Storm* (Boston: Houghton Mifflin, 1948), p. 40.
6 See Kalevi. J Holsti, *International Politics: A Framework for Analysis* (Englewood Cliffs, New Jersey, USA: Prentice Hall, 1983).
7 John Spanier, *Games Nations Play* (New York: Holt, Rinehart and Winston, 1984), pp. 163–4.

8 See Richard Connaughton, *A Brief History of Modern Warfare: The True Story of Conflict from The Falklands to Afghanistan* (London: Constable and Robinson Ltd, 2008).

9 Ministry of Defence, *British Defence Doctrine*, Joint Warfare Publication 0-01, (London: MoD, 1996) p. 1.8.

10 Ibid. pp. 6.4–6.14.

11 Figures are drawn from, Charles Heyman (ed.), *The Armed Forces of the United Kingdom 2010–11* (Barnsley: Pen and Sword, 2009).

12 HM Government, *Securing Britain in an Age of Austerity: the Strategic Defence and Security Review (SDSR)*, CM7948 (London: The Stationery Office, 2010).

13 House of Commons Defence Select Committee Report, *The Strategic Defence and Security Review and the National Security Strategy*, HC 761 (London: The Stationery Office, August 2011).

14 Defence Committee: Select Committee Announcement, 1 August 2011, 'Wide Ranging Concern about Strategic Defence and Security Review' (Defence Select Committee, 2011).

15 National Audit Office, *Ministry of Defence: Major Projects Report 2009* (London: The Stationery Office, December 2009).

16 This estimate is based on the assumption that the defence budget under a Labour government from 2010 would have remained level in real terms, compared against the forecast budget needed to finance MoD known forward plans from 2010 to 2020/2021. See Malcolm Chalmers, 'Capability Cost Trends: Implications for the Defence Review,' *Future Defence Review Working Paper* No. 5 (London: RUSI, January 2010).

17 Lord Levene has conducted a wide-ranging review of the Ministry of Defence.

18 The seven defence tasks are: the defence of the UK and overseas territories, the provision of strategic intelligence, the provision of nuclear deterrence, support to civil emergencies, the projection of power through expeditionary interventions, the provision of defence contribution to UK influence, and the provision of security for stabilisation.

19 Ministry of Defence, *National Security Through Technology: Technology, Equipment and Support for UK Defence and Security*, CM8278 (London: The Stationery Office, February 2012). See also, Ministry of Defence, *Defence Green Paper – Equipment, Support and Technology for UK Defence and Security*, CM7989 (London: The Stationery Office, 2010).

20 Ministry of Defence, DASA (Defence Expenditure Analysis data) www.dasa.mod.uk/modintranet/ukds/ukds2010, accessed on 16 June 2011.

21 Private interview, 1 June 2011.

22 HM Government, *National Security Strategy and Strategic Defence and Security Review 2015: A Secure and Prosperous United Kingdom*, CM9161 (London: The Stationery Office, 2015), para 1.1, p. 9.

23 HM Government, *Securing Britain in an Age of Uncertainty: The Strategic Defence and Security Review*, CM7948 (London: The Stationary Office, 2010).

24 Ibid., p. 2.

25 Ibid.

26 HM Government (2015), CM9161, op. cit.

27 Ministry of Defence, *SDSR 2015 Key Facts* (London: Ministry of Defence, 2015) www.gov.uk/government/uploads/system/uploads/attachment_data/file/494895/SDSR_2015_Booklet_vers_15.pdf.

28 HM Government (2015), CM9161, op. cit., p. 45 and p. 82.

29 The US Third Offset Strategy is an American attempt to find a technology or equipment that will deliver a competitive edge for the US armed forces that provides a margin of victory against technologically adept and disruptive adversaries. John Louth and Trevor Taylor, 'The US Third Offset Strategy: Hegemony and Dependency in the

Twenty-First Century', *RUSI Journal*, June 2016. https://rusi.org/publication/rusi-journal/us-third-offset-strategy-hegemony-and-dependency-twenty-first-century.
30 Richard Barrons and Ewan Lawson, 'Warfare in the Information Age', *RUSI Journal*, 2016, 161(5).
31 Michael Clarke, 'Reacting to the 2015 Strategic Defence and Security Review', RUSI Commentary, 23 November 2015, https://rusi.org/node/15250, accessed 15 May 2017.

Bibliography

Barrons, R and Lawson, E (2016), 'Warfare in the Information Age', *RUSI Journal*, 161(5).

Chalmers, M (2010), 'Capability Cost Trends: Implications for the Defence Review,' *Future Defence Review Working Paper* No. 5 (London: RUSI, January 2010).

Churchill, WS (1948), *The Second World War, The Gathering Storm* (Boston: Houghton Mifflin).

Clarke, M (2015), 'Reacting to the 2015 Strategic Defence and Security Review', RUSI Commentary, 23 November 2015, https://rusi.org/node/15250, accessed 15 May 2017.

Connaughton, R (2008), *A Brief History of Modern Warfare: The True Story of Conflict from The Falklands to Afghanistan* (London: Constable and Robinson Ltd).

Defence Committee (2011): Select Committee Announcement, 1 August 2011, 'Wide Ranging Concern about Strategic Defence and Security Review' (Defence Select Committee).

Harari, YN (2015), *Homo Deus: A Brief History of Tomorrow* (London: Harvill Secker).

Heyman, C (ed, 2009), *The Armed Forces of the United Kingdom 2010–11* (Barnsley: pen and Sword).

HM Government (2010), *Securing Britain in an Age of Austerity: the Strategic Defence and Security Review* (SDSR), CM7948 (London: The Stationery Office).

HM Government (2010), *Securing Britain in an Age of Uncertainty: The Strategic Defence and Security Review*, CM7948 (London: The Stationary Office).

HM Government (2015), *National Security Strategy and Strategic Defence and Security Review 2015: A Secure and Prosperous United Kingdom*, CM9161 (London: The Stationery Office).

Holsti, KJ (1983), *International Politics: A Framework for Analysis* (Englewood Cliffs, New Jersey, USA: Prentice Hall).

House of Commons Defence Select Committee Report (2011), *The Strategic Defence and Security Review and the National Security Strategy*, HC 761 (London: The Stationery Office, August 2011).

Louth, J and Quentin, P (2014), 'Making the Whole Force Concept a Reality', *RUSI Briefing Paper*.

Louth, J and Taylor, T (2016), 'The US Third Offset Strategy: Hegemony and Dependency in the Twenty-First Century', *RUSI Journal*, June 2016. https://rusi.org/publication/rusi-journal/us-third-offset-strategy-hegemony-and-dependency-twenty-first-century.

Ministry of Defence (1996), *British Defence Doctrine*, Joint Warfare Publication 0-01, (London: MoD).

Ministry of Defence (2010), *Defence Green Paper – Equipment, Support and Technology for UK Defence and Security*, CM7989 (London: The Stationery Office).

Ministry of Defence (2010), Defence Expenditure Analysis data, www.dasa.mod.uk/mod-intranet/ukds/ukds2010, accessed on 16 June 2011.

Ministry of Defence (2012), *National Security Through Technology: Technology, Equipment and Support for UK Defence and Security*, CM8278 (London: The Stationery Office, February 2012).

Ministry of Defence (2015), SDSR 2015 Key Facts (London: Ministry of Defence). www. gov.uk/government/uploads/system/uploads/attachment_data/file/494895/SDSR_ 2015_Booklet_vers_15.pdf.

National Audit Office (2009), *Ministry of Defence: Major Projects Report 2009* (London: The Stationery Office, December 2009).

Ordnance Survey Conference (2017) – the Cambridge Futures' Conference – held at Keble College, Oxford on 3 July 2017.

Pinker, S (2011), *The Better Angels of our Nature: Why Violence has Declined* (New York: Viking).

Spanier, J (1984), *Games Nations Play* (New York: Holt, Rinehart and Winston).

The World Health Organisation (2012), 'Global Health Observatory Data Repository'.

UK Ministry of Defence (2015), *SDSR 2015 Key Facts* (London: Ministry of Defence), www.gov.uk/government/uploads/system/uploads/attachment_data/file/494895/ SDSR_2015_Booklet_vers_15.pdf.

11 Conclusion

Defence practice – from analogue to digital? The Defence Extended Enterprise

Introduction

This book has been about the story of UK defence in the twenty-first century. Deliberately, we highlighted the military last in Chapter 10, as modern defence is a complicated and nuanced public–private practice that is much about the armed forces, but also has the armed forces as just one component of defence capability. So that defence as:

- Policy and politics, Chapter 2;
- Management, Chapter 3;
- Technological development, Chapter 4;
- Industrial policy, Chapter 5;
- Exports and engagement, Chapter 6;
- Skills and competencies, Chapter 7;
- Community action, Chapter 8;
- Teamwork and partnering, Chapter 9; and
- The military, Chapter 10;

can all be conceptualised as the Defence Extended Enterprise. As we have seen, military personnel, civil servants, reserves, contractors, and myriad elements of multiple supply chains merge to deliver sought military effects. This 'Whole Force' approach or 'enterprise' management stance is how the UK will exercise defence and security in the years ahead. The components of this enterprise need to be mapped and managed effectively if defence is to be assured. We believe this is the elemental challenge facing decision-makers today.

So, within the Ministry of Defence, those decision-makers have to decide whether they are consumers or producers of defence capabilities. How defence as an enterprise answers this question is its 'analogue or digital' moment.

The MoD – consumer or producer?

Within much of the business literature there is a clear distinction between commercial entities which are buyers of goods or services – therefore meeting a need or selling-on into the marketplace – and those which are manufacturing entities that blend raw materials, technologies, production know-how, and specific components to generate and, thereafter, support a particular product or service.[1] Based on this distinction, a first step for defence analysts is to consider whether the MoD should be seen primarily as a buyer and user of goods and services, akin to the individual consumer, or whether it should be viewed in the guise of a major entity that sources, blends, amends, develops, and delivers military products or effects.

When the individual consumer opts to buy a television or even selects a builder for a home extension, he/she does not normally worry about the impact of the choice on supply chains or even how the television manufacturer selects its suppliers. The focus is on the price, performance, and reliability of the final product and the consumer holds the brand manufacturer responsible for all the features of the product, regardless of where they came from: should the Toyota owner ever press the door opener on the key fob and find the car stayed locked, he or she would not observe that Toyota must have chosen a poor lock supplier.

Indeed, some officials and political figures associated with defence view the MoD as essentially a purchaser of goods and services from the private sector for deployment to meet defence and security ambitions. In interviews with the authors, MoD officials and service personnel took this initial instinctive position.[2] A contrasting view underlines that the MoD is responsible for 'producing' things and not just 'using' things. The MoD's own reporting and performance-measurement systems show that it is responsible for the generation of outputs (force elements able to act at varying degrees of notice) and outcomes (deterrence and success on operations).[3] The MoD's central task is to produce UK defence policy and then to direct the generation of military capabilities that support the delivery of that policy. In the event of a government decision to use those capabilities on an operation, it is then the role of the MoD to oversee and even control that use so that it supports political objectives and operates within government-specified constraints.

Under this approach, the MoD, and the armed forces and agencies within it are significantly analogous to manufacturing organisations, bringing together all the diverse elements required for usable and sustainable defence capability. Some of those elements they generate within the governmental defence sector, while (many) others are sourced from outside.

It would be difficult to exaggerate the dependence of UK defence on suppliers in the private sector, where the MoD spends more than half of its money. Most obviously, the MoD obtains all its equipment, spare parts, and fuels from the private sector, with the days of the state arsenal being long gone. Moreover, the MoD is remarkably capital intensive, investing around a quarter of its budget each year in the development and production of new equipment. While the valuation of military equipment is clearly problematic in many ways, it is striking that the MoD's 2014–2015 Annual Report records the MoD holding property, plant, and equipment assets worth £95 billion and intangible assets – mainly reflecting research and development (R&D) spending on equipment – of

£25.5 billion,[4] which together represent more than three times the MoD's annual cash spend.

Within this context, a major influence on defence policy has been the strand of 'New Public Management'[5] thought which asserts that government bodies are generally not as good at the generation of goods and services as the private sector. This, alongside a commitment and preference to reduce the number of civil servants, has left the MoD predisposed to further outsourcing, not least of services. In 2015 the MoD sold the Defence Support Group, formerly an MoD agency, to Babcock, prompting the Think Defence website to establish a review entitled 'What is There Left to Outsource?'[6]

While less significant in overall financial terms, it is striking how much the MoD has passed responsibility for even the generation of innovative ideas and initiatives for change to the private sector. For example, in 2015 in the Defence Equipment and Support (DE&S) organisation, Bechtel and CH2M Hill were advising on project- and programme-management transformation while PwC was advising on human resources. Even when the MoD sought to manage aspects of the extended enterprise, it needed others to help provide guidance on how to do it. Below is an excerpt from the website of PA Consulting:[7]

> The UK Ministry of Defence (MOD) submarine programme enables the UK to maintain a continuous 'at sea' strategic deterrence and is one of the world's most complex Programmes. The challenge for the MOD is to deliver the programme to time and cost through four sole providers of sub-marine build, propulsion, support and warhead capabilities, for elements of which the UK can be the only customer.
>
> Without being able to apply normal commercial levers, the MOD saw that managing these suppliers tactically on a project-by-project basis would not be enough. As such, the MOD first asked PA to work with them to design a new strategic supplier management capability. PA led work as part of a joint team to establish this capability through the long-term Submarine Support Partner arrangement …
>
> [We] established governance structures, supplier strategies, performance monitoring mechanisms and analysis capabilities that have started to deliver results.

The conclusion is clear and overt from a commercial perspective such as this: without external expertise, the MoD would be unable to meet its obligations to the nation.

The MoD must be able to both generate capability and then, when called upon, to use it. This means that the MoD requires significant flexibility and agility from its supply chains: defence equipment used on operations generally requires more fuel, spare parts, and so on than when it is in a training or standby role. Moreover, many operations since the end of the Cold War (and before that, the Falklands War) had not been envisaged in long-term defence plans, con-tained an element of surprise, and thus placed special demands on defence suppliers to accelerate production of certain items, to modify equipment for a particular campaign and theatre, and then to come up with novel products.[8] The term Urgent Operational Requirement has become familiar in defence discourse, with UK forces having generated hundreds of them since the end of Cold War.

The MoD can place significant demands on its suppliers, requiring them first to be efficient and effective in delivering their goods and services, and then flexible and agile in terms of being responsive to radical and rapidly changing circumstances. Finally, in defence, where human lives, as well as operations of crucial importance to governments, can be at stake, it should be hoped that supplies to the MoD should be resilient – that is, not excessively vulnerable to shock. Resilience has three dimensions: the capacity to avoid disruptive shock; the mitigation of an immediate response should such a shock occur; and the rapid, as opposed to protracted, recovery from shock.

Managing and integrating the defence supply base

If there are parallels between the MoD and large manufacturing organisations, how do such bodies behave with regard to their supply base and what guidance can be derived from contemporary management thought?

Very often, large companies have developed a significant external supply base and focused their internal efforts on a few specific areas. In the commercial world, the concept of core competencies has had a huge influence on the shape of organisations.[9] The idea is that organisations, and their component parts, cannot be expected to excel at more than a few things and that therefore they should focus on what they absolutely can be good at and need to be good at. Things that they need but do not excel in producing should be obtained from those that specialise in those areas. The result has been extensive outsourcing of both support functions, such as catering and facilities management, and the supply of goods and services needed directly for a company's products; in contrast to the days when Ford sought to run its own rubber plantations to supply the raw material for its tyres, Ford no longer specialises in tyre technology but relies on others who do.[10] Nike focuses on design and marketing, and outsources actual manufacture, as do many clothing firms. Apple addresses technological innovation, design, and product development, but oversees manufacturing rather than doing much itself. In his influential book first published in 1992, Martin Christopher argued that it was more useful to measure competition among supply chains than simply among lead manufacturers.[11]

In addition, the transport and communications possibilities offered by developing relevant technologies, and the political embrace of more open international trade agreements, have meant that supply chains have become complex and global for many manufactured goods. The components and sub-systems of many familiar products such as automobiles and personal computers are sourced from around the world, and the most complex technologies are now just components within a larger platform, sourced and traded globally.

These trends have significantly increased the importance and domain of 'supply-chain management' as enterprises recognise the need to be aware of not just the current performance but also the medium- and long-term health of those organisations on which they depend. Volkswagen, with 6,000 suppliers in Europe alone, notes that, not least because of the rate of technological change, '[in] future, a key success factor will be a highly efficient global supplier network', and has launched its Future Automotive Supply Tracks (FAST) programme.[12] Companies need also to be sensitive as to what should be the limits of outsourcing: Boeing, for instance, was criticised for having outsourced too much on its 787 project. Also, as both the Airbus A380 and the Boeing 787 experienced delays in entering service and large-scale production, the two prime contractors had to monitor key suppliers lest some went out of business as a result of anticipated revenues from aircraft production not being available.[13]

Reflecting such developments, academic analysts articulated the concept of the 'extended enterprise', which comprises the lead company and all those on whom it depends.[14] Supply-chain management emerged as a much broader topic than checking whether a firm delivered to the right place, at the right time, to the agreed quality for the agreed price. The best firms looked to the future development of their suppliers, helping them to innovate and listening to their ideas. There are two key dimensions of supply-chain management, of which 'understanding' is the first: purchasers need a lot of information about the current and likely future condition of suppliers that are important to them. Precisely which information should be collected depends on market structures and dynamics so there is no set database for a purchaser to populate. For example, the supply-chain information needs of a company such as BAE Systems – which relies on hi-tech components and, sometimes, volatile raw materials – are significantly different from a firm such as Aldi, the high-street discount store.

Second is the action dimension: purchasers have to select what they are going to do to incentivise and support their suppliers to ensure their survival and, preferably, continual improvement. Moreover, the dominant relationship type in the extended enterprise has been collaborative rather than adversarial.[15] The objective of the Volkswagen FAST programme noted above is 'to prioritise investments and make efficient use of resources *in closer cooperation with suppliers*' (authors' emphasis).[16]

It also became clear that risk must be a significant consideration in a firm's decision to rely on complex and global supply chains, as experience showed that misjudgement or neglect of supply-chain management could be catastrophic.[17] There are a number of high-profile cases where firms were severely impacted by supply-chain failure: in 1997, Toyota car production in Japan was halted for several days by a fire in a supplier's small factory producing brake and clutch parts;[18] in 2000, the Ericsson mobile-phone business suffered fatal damage from a small fire in a Philips factory in Albuquerque, New Mexico;[19] and the 9/11 attacks had repercussions well beyond the firms based in the World Trade Center.[20] Thus, leading businesses were drawn to give attention to risk, resilience (the ability to bounce back from shock and damage), and agility throughout the extended enterprise rather than just within their own organisations.

From this short analysis it can be gleaned that four key factors for supply-chain resilience have been asserted: visibility (of the supply chain); collaboration (among the organisations involved); flexibility (the capacity to adapt); and control (mechanisms to ensure that policies are followed).[21] Clearly, though, it is not enough to simply recognise the existence of the extended enterprise. There is a need also actively to manage it by anticipating the future; giving direction; and planning, organising, co-ordinating, and monitoring performance.[22] Moreover, responsibility for the extended enterprise must lie with a specific person and group: in a private company that would be the chief executive working with the board. Finally, extended-enterprise management seems to pay off: Theano Lia-kopoulou of McKinsey told a London audience in May 2015 that those who develop and collaborate well with their suppliers create 15 times the value of the average relationship.[23] While recognising that such a sum is hard to quantify, Oxford Economics concluded in an analysis of BAE Systems' contribution to the UK economy that collaborative supplier relationships were critical to the company's success.[24]

So, the MoD and, more broadly, the government, must manage the components of defence as an extended enterprise, planning, encouraging, rewarding, and, where necessary, reprimanding public and private sector contributors – the building blocks of our capability and security. This is a complicated and difficult thing to do but, given the analysis within this book, at least decision-makers might know the task that confronts them.

Notes

1 For example, see Mark Saunders, Philip Lewis and Adrian Thornhill, *Research Methods for Business Students*, 2nd edition (Harlow: Pearson, 2000).

2 The authors conducted a number of semi-structured interviews with Ministry of Defence (MoD) officials and members of the armed forces between May and September 2015.

3 See multiple annual editions of the MoD's Annual Report and Accounts. For the 2014–2015 edition, see MoD, *Annual Report and Accounts: 2014–2015*, HC 32 (London: The Stationery Office, 2015).

4 MoD, *Annual Report and Accounts: 2014–2015*.

5 New Public Management developed in the 1980s and 1990s as a way of championing commercial ideas and practices over what was traditionally held to be 'public.' See Tony Tinker and Tony Puxty, *Policing Accounting Knowledge* (London: Paul Chapman Publishing, 1995).

6 Think Defence, 'What is There Left to Outsource', 4 July 2015, www.thinkdefence. co.uk/2015/07/what-is-there-left-to-outsource/, accessed 25 August 2015.

7 PA Consulting, 'Enhancing the MOD's Capability to Manage "Monopolistic" Suppliers on a £27 Billion Programme', www.paconsulting.com/our-experience/ enhancing-the-mods-capability-to-manage-monopolistic-suppliers-on-a-27-billion-programme-uk-ministry-of-defence/, accessed 8 June 2015.

8 For illustrations of UK experiences in mobilising the private sector to support military operations since 1991, see Trevor Taylor, John Louth, and Henrik Heidenkamp, 'Industry and the Military Instrument', in Adrian L Johnson (ed.), *Wars in Peace: British Military Operations since 1991* (London: RUSI, 2014), pp. 291–320.

9 Arguably, the ground-breaking publication here is Gary Hamel and CK Prahalad, *Breakthrough Strategies for Seizing Control of Your Industry and Creating the Markets of Tomorrow* (Boston, MA: Harvard Business School, 1994).

10 See Michael E Porter, *Competitive Strategy: Techniques for Analyzing Industries and Competitors* (New York: Free Press, 1980).

11 The book has been regularly updated. See Martin Christopher, *Logistics and Supply Chain Management*, 4th edition (London: Financial Times, 2010).

12 Comments made by Francisco Javier Garcia Sanz, Volkswagen board member responsible for procurement, as reported in Supply Chain 24/7, 'VW Ready to Transform Automotive Supply Chains', 3 March 2015, www.supplychain247.com/article/vw_ready_to_transform_automotive_supply_chains, accessed 3 September 2015.

13 'Boeing also had to pay strategic partners compensation for potential profit losses stemming from the delays in production', for this and wider Boeing problems with supply-chain management and the 787, see Steve Denning, 'What Went Wrong at Boeing?', *Forbes*, 21 January 2013. See also Yasuhiro Monden, *Toyota Production System: An Integrated Approach to Just-in-Time*, 3rd edition (Norcross: Engineering and Management Press, 1998).

14 See James E Post, Lee E Preston and Sybille Sachs, 'Managing the Extended Enterprise: The New Stakeholder View', *California Management Review* (Vol. 45, No. 1, Fall 2002); Dan Jones and Jim Womack, *Seeing the Whole: Mapping the Extended Value Stream* (Cambridge, MA: The Lean Enterprise Institute, 2002); Andrew Humphries and Richard Gibbs, *Enterprise Relationship Management: A Paradigm for Alliance Success*, new edition (Farnham: Gower, 2015).

15 See Jeffrey H Dyer, *Collaborative Advantage: Winning through Extended Enterprise Supplier Networks* (Oxford: Oxford University Press, 2000); Ed Davis and Robert Spekman, *The Extended Enterprise: Gaining Competitive Supply Chains through Collaborative Supply Chains* (Upper Saddle River, NJ: Financial Times, 2004).

16 Supply Chain 24/7, 'VW Ready to Transform Automotive Supply Chains'.

17 See Steve Denning, 'The Boeing Debacle: Seven Lessons Every CEO Must Learn', *Forbes*, 17 January 2013; Donald Waters, *Supply Chain Risk Management: Vulnerability and Resilience in Logistics*, 2nd edition (London: Kogan Page, 2011); Institute of Risk Management, 'Extended Enterprise: Managing Risk in Complex 21st Century Organisations', 2014.

18 Valerie Reitman, 'Toyota Halts Japan Production after Fire Destroys Factory', *Wall Street Journal*, 3 February 1997.

19 See Amit S Mukherjee, *The Spider's Strategy: Creating Networks to Avert Crisis, Create Change, and Really Get Ahead* (New York, NY: FT Press, 2008), chapter 1; Yossi Sheffi, *The Resilient Enterprise: Overcoming Vulnerability for Competitive Advantage* (Cambridge, MA: MIT Press, 2007).

20 According to the *New York Times*, the damage to business caused by the attacks amounted to $123 billion and the total cost including the direct damage and the consequent increased defence and security spending amounted to $3.3 trillion. See Shan Carter and Amanda Cox, 'One 9/11 Tally: $3.3 Trillion', *New York Times*, 8 September 2011.

21 Kelly Marchese, Siva Paramasivam and Michael Held, 'Bouncing Back: Supply Chain Risk Management Lessons from Post-Tsunami Japan', Industry Week, 9 March 2012, www.industryweek.com/global-economy/bouncing-back-supply-chain-risk-management-lessons-post-tsunami-japan, accessed 24 August 2015.

22 These basic management functions were first articulated in 1916 in Henri Fayol, *Administration industrielle et générale* (Paris: Dunod, 1916).
23 Paul Snell, 'Four Factors that Define Procurement Excellence – Expert', *Supply Management*, 8 May 2015.
24 Oxford Economics, 'The Economic Contribution of BAE Systems to the UK in 2009', 2011.

Index

Page numbers in **bold** denote tables, those in *italics* denote figures.